# Astronomers' Universe

More information ab  er.com/series/6960

Pierre Léna

# Racing the Moon's Shadow with Concorde 001

Pierre Léna
Université Paris Diderot & Observatoire de Paris
Paris
France

Translation by Stephen Lyle



Original edition: *Concorde 001 et l'ombre de la Lune*. © Editions Le Pommier - Paris, 2014

ISSN 1614-659X ISSN 2197-6651 (electronic)
Astronomers' Universe
ISBN 978-3-319-21728-4 ISBN 978-3-319-21729-1 (eBook)
DOI 10.1007/978-3-319-21729-1

Library of Congress Control Number: 2015955559

Springer Cham Heidelberg New York Dordrecht London


*Cover illustration and Frontispiece*: Racing the Moon by Don Connolly, Sydenham, Ontario, Canada. Conception by L. Robert Morris, Ottawa, Canada. Composition by Don Connolly and L. Robert Morris. Acrylic on board (49x68 cm), 2004. This artist view is done with the assumption that the viewer is located 3,000 m below the aircraft, at latitude 16.19°N, longitude 14.38°E, altitude of Concorde 001 17,602 m above Niger, time : 12h 07 min 24 s UTC.

Springer International Publishing AG Switzerland is part of Springer Science+Business Media (www.springer.com)

*I dedicate the English translation of this book to the young students of Ross School, where they happily discover the beauty of science, the richness of human cultures, and how best to serve their sisters and brothers, inhabitants of our planet*

# Acknowledgements

I am grateful to many people who have helped me to relive the details of this story: first and foremost, André Turcat, for our friendly discussions and for allowing me to use his own recollections, but also Michel Rétif, our flight mechanic; Hubert Guyonnet, our radio navigator, and Henri Perrier, sadly missed; my astronomer colleagues, John Beckman, Donald Hall, Donald Liebenberg, Alain Soufflot, Paul Wraight, and Serge Koutchmy, who shared our enthusiasm for the various science programmes on the flight of 30 June 1973; the team at the *Musée de l'air et de l'espace* in Le Bourget, including the director Catherine Maunoury and the curator Christian Tilatti, for setting up a permanent exhibition beside Concorde 001; Jean-Pierre Sarmant, for careful and expert copy-editing, and the anonymous referee at the *Centre national du livre*; Robert L. Morris, a Canadian university teacher, as keen on Concorde as he is on eclipses, who provided me with many documents and ideas for this work. The Canadian artist Donald Connolly went to great lengths to describe the context of the 1912 and 1973 eclipses and generously allowed us to reproduce his two paintings here. The illustrations owe much to the photographer Jean Mouette and the philatelist Henri Aubry, and also to Vincent Coudé du Foresto, who flew aboard the Air France Concorde for the 1999 eclipse, André Girard, who led the environmental tests on Concorde from the *Office national d'études et de recherches aéronautiques*, and Jim Lesurf, who was part of the British science team. Thanks must also go to the computer expert and amateur astronomer Xavier Jubier.

The writing of the book was made possible through the generous support of the *Fondation des Treilles*,[1] where I was able to stay for a study visit during which Emmanuelle and Valérie provided invaluable assistance. Finally, I am extremely grateful to my French publisher Sophie Bancquart for her unfailing trust in me.

The publication of the English translation was made possible through the generosity of Mrs. Courtney Ross. The author would like to express his warmest gratitude to her, and to thank the translator Stephen Lyle, as well as the staff of the publisher Springer.

[1] The *Fondation des Treilles* was set up by Anne Gruner-Schlumberger with the aim of encouraging a dialogue between science and the arts and hence to stimulate progress in contemporary research and creativity. It hosts researchers and writers at the *Domaine des Treilles* in the Var region of France: http://www.les-treilles.com

# Contents

# Overture

[...]*mon luth constellé*
*Porte le Soleil noir* [...]
Gérard de Nerval[2]

On that 30 June 1973, the summer Sun rose over Las Palmas, the capital of the Canary Islands off the African coast, but this time there was something different in its appearance. A piece of the solar disk was missing, blacked out by the edge of the Moon, for our satellite had just begun to move between the Sun and the Earth. At sunrise on the same day, far to the west in Dutch Guiana (now Surinam), the tree frogs of the species *Hyla calcarata* began to chorus, despite the unusual time of day.[3] For indeed it was there on the equator that the great dark disk formed by the Moon's shadow on the surface of the Earth, surrounded by penumbra, began its race towards the east, moving at a speed of more than 2000 km an hour relative to the ground, soon to cross the African coastline. Astronomers, who know how to predict not just the occurrence, but all the details of these eclipses, had explained how this one would be total at every point on the Earth that the shadow would sweep past and partial in each region falling only in the penumbra. Better still, in Africa, total obscurity would last, exceptionally, for 7 min, and the sky would be so dark that the stars would be visible even at midday. This record length would make it the eclipse of the century, as announced by the media. By around 10 o'clock in the morning, the black indentation had grown to block out almost half of the Sun's disk. Slowly, a great white bird began to move across the tarmac at Las Palmas airport to position itself on the runway. Fitted out with a bright red survival suit—standard dress for test flight crew members, in case they need to be fished out of the sea—I found myself aboard Concorde 001F-WTSS, the prototype of the future supersonic

[2] [...]*and my constellated lute*
*Bears the black Sun* [...]

[3] Jean Lescure: *Comportement vocal des amphibiens et des oiseaux au Surinam*. In: *Soleil est mort: l'éclipse totale du 30 juin 1973*, G. Francillon & P. Menget (eds), pp. 91–103 (1979). See the bibliography.

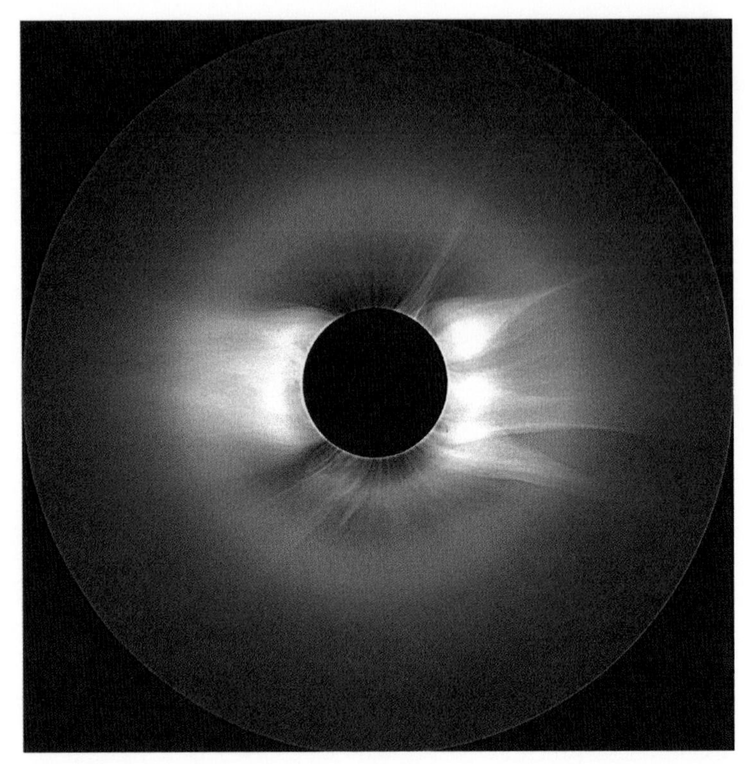

**Fig. 1** The solar corona, photographed by Serge Koutchmy from Moussoro in Chad during the total eclipse of 30 June 1973. The observer is in the umbra (*full shadow*), 100 km *south* of the line of centrality. © Institut d'astrophysique de Paris/CNRS

passenger aircraft to be flown by Air France and British Airways, with seven other astronomers. At the controls, André Turcat, the famous pilot who had directed all the test flights of this plane since 1969, and Jean Dabos both kitted out with the same survival gear. The four other men making up the crew were ready at their posts: "Sierra Sierra, ready for takeoff", announced the control tower. Exactly on time within a second, the beautiful aircraft took off into the trade winds, accompanied by the roar of its four jet engines, flying to meet the shadow of the Moon somewhere above the barren deserts of Mauritania. Less than an hour later, having accelerated to more than twice the speed of sound in the stratosphere, to which it rises in order to attain its cruising altitude, Concorde 001 was about to encounter the Moon's shadow at precisely the place and time identified by our calculations. The accuracy of the rendezvous was extraordinary, since we reached it within 1 s and at less than 2 km from the ideal point. The plane's night-time navigation lights went on even though it was close to midday local time and the Sun was close to our zenith. Flying in the lunar shadow, which was moving at the same speed as us, the plane would remain in total darkness for 74 very long minutes, while each of the astronomers was getting busy with the instruments they had brought along to study the Sun and its corona, taking advantage of this unique opportunity to observe for such a long period of time. When we landed at Fort Lamy, today N'Djamena, the capital of Chad in the very heart of the African continent, I was filled with emotion. This dream which, as a young astrophysicist and teacher, I had sketched out barely a year earlier had just been made a reality. No man had ever seen the Sun eclipsed for such a long time, no flight crew had ever carried out such a difficult encounter so faultlessly, no plane had ever provided such a fine observatory for its team of awestruck astronomers. Forty years on, this record remains unbeaten.

**Fig. 2** The shadow of the Moon on the Earth, surrounded by the penumbra. The photograph was taken from space by the French astronaut Jean-Pierre Haigneré on board the Mir space station during the total eclipse of the Sun of 10 August 1999, which was visible in France. © CNES/JP Haigneré, 1999

In the Mauritanian sky, as Tuaregs gazed astonished upon this Sun so quick to hide, three parallel stories would finally come together. The first began with humanity which, in all its cultures, had long feared the total eclipse of the Sun, but then sought to understand its cause, and finally to use it scientifically to study the celestial body which brings us light, heat, and life. The second story concerns the extraordinary aircraft that was Concorde, and for which 001 was the first prototype, flying since 1969 and based in Toulouse. So how was this plane deflected from its main objective, that is, the flight tests of a future supersonic passenger plane, to become a scientific laboratory chasing after the shadow of the Moon? Finally, the third story concerns me more directly, since it took me years to become a research scientist, to discover my interest for the Sun and infrared light, to learn the difficult techniques required to fly telescopes aboard aircraft, to build a team that could accompany me in this project, and to dream up this African flight and turn it into a reality.

**Fig. 3** At 10 h 08 Universal Time on 30 June 1973, the prototype Concorde 001, registration F-WTSS, took off from Las Palmas airport in the Canary Islands to meet up, somewhere over West Africa, with the Moon's shadow, already slipping rapidly over the Earth's surface. The afterburn (or reheat) of the four engines has been activated to get the 136 tonne airliner off the ground. © Jim Lesurf

This is the encounter that I would like to narrate, and it is a true story. The telescopes and the Sun, the supersonic aircraft chasing the shadow, the astronomers, the engineers, and the pilots who set this exceptional record really did exist. This was no video game in which the player travels at will through space and time, nor a comic strip with something fantastical in every picture. It was a collective human adventure which we really experienced. Here I wish to transmit this tale to those who lived through those times, but also to the younger generation in the hope of kindling other dreams and helping them to confront the hard realities of life. I would so much like to encourage them to aim high and excel themselves in new adventures, fulfilling those dreams. But I do have one regret, for my female readers. There are so few women in this story: astronomers, engineers, and crew were all men, and there is nothing I can do about that. However, during my working life, I have had brilliant female colleagues, an honour to astrophysics, and many of my former students have acquired international acclaim, such as Nabila Aghanim, whom I met in Algiers many years ago, now exploring the depths of the past universe. Aviation also has its heroines, such as Commander Caroline Aigle, former student at the *Polytechnique*, later fighter pilot, who sadly passed away in 2007, or Catherine Maunoury, twice winner of the World Aerobatic Championships and today director of the *Musée de l'Air et de l'Espace* in Le Bourget (France), which has housed Concorde 001 since 1973. And not forgetting space, with Claudie Haigneré, a doctor who decided to become an astronaut and spent a total of more than

25 days in space during her various missions. Women are as much needed in science as in exploits of bravery and intellect.

Many glorious and fruitful projects have marked the twentieth century, such as those undertaken by the men and women who went into space, and in particular the Apollo missions to the Moon. They mobilised so many more people, and much greater resources and intelligence than our modest eclipse expedition, and they brought back immeasurably greater scientific results. However, in my own adventure, astronomy and aviation came together in a quite exceptional, and perhaps unique, way that may not be rivalled for some time to come. So I feel the story deserves to be told.

But why did I wait 40 years for this? After Concorde, my existence was occupied by so many other astronomy projects, leaving me little time to look back. Today, before all those involved should pass away, it seems the right moment to recount all this to the younger generation.

NOTE: The reader will find several supplementary explanations of a scientific nature in appendix at the end of the book, but these are not essential to follow the story.

# Chapter 1
# Eclipses and Humankind

## 500 Million Years in the Sea

The tiny but elegant cephalopod called nautilus[1] has a calcium carbonate spiral shell which develops day by day with the addition of a new growth line. For at least 500 million years, nautili have come up to the surface on the day of the full Moon to spawn, at which point they insert a partition or septum in the series of growth lines. Taking a fossil nautilus, dated using geological techniques, and counting the number of growth lines separating two successive septa, we obtain the duration of the lunar month at the time when that particular nautilus was alive.[2] We thus discover that this duration was shorter than it is today. Indeed, the Moon is slowly moving away from the Earth, and this increases its period of rotation about its host planet. The distance is increasing at an average rate of 38 m per century, as can be measured very accurately today using laser ranging. This has been brought about by the dissipation of energy produced by friction in the tides on Earth. The tides are caused by the gravitational pull of the Moon (and also the Sun) on the Earth's oceans. The considerable kinetic energy involved in the motion of these huge masses of water is extracted from the gravitational energy associated with the masses of the Earth and the Moon. As the frictional effects accompanying the motion of the water finally transform part of this energy into a rise in the ocean temperature, at the expense of the Earth–Moon pair, these bodies gradually move apart. But when the Moon moves away from the Earth, its apparent size in the sky will decrease. It is thus thanks to a happy coincidence that, since historical times, the lunar and solar disks viewed from the Earth's surface have practically the same apparent size when we look at them in the sky.

When it happens that the Moon, during its orbit around the Earth, passes between the Sun and a point on the Earth's surface, its disk will exactly cover the solar disk

[1] http://en.wikipedia.org/wiki/Nautilus/

[2] http:/www.thegreatbetween.com/the-chambered-nautilus/



P. Léna, *Racing the Moon's Shadow with Concorde 001*, Astronomers' Universe,
DOI 10.1007/978-3-319-21729-1_1

and a total eclipse of the Sun will occur. But in a few tens of thousands of years, such a precise obstruction of the Sun's light will no longer be possible, so we must take full advantage. Furthermore, if the Moon were to follow a precisely circular orbit around the Earth, then all eclipses would look exactly alike. But its orbit is in fact an ellipse and its distance from the Earth thus varies slightly during the lunar month: sometimes the eclipse will thus be annular, leaving a thin bright ring of the solar surface (the diamond ring) in view, and sometimes it will be total, when the Earth—Moon separation is minimal.

Let us be more precise for a moment. The distance from the Earth to the Sun also changes slightly, since likewise, the Earth's orbit is not quite a circle but another ellipse, even though very close to circular in this case. The apparent diameter of the solar disk as it appears to us in the sky thus varies between $32'35''$ and $31'31''$, where a single prime indicates 1 min of angle, or the sixtieth part of a degree, and the double prime indicates 1 s of angle. The apparent diameter of the Moon, also on its elliptical orbit around the Earth, varies between $33'1''$ and $29'22''$. We see therefore that, depending on the positions of the Earth and Moon on their elliptical trajectories at the exact time when the eclipse occurs, the lunar disk will obscure the solar disk to differing extents. And apart from these variable distances, many other factors are relevant to determining the circumstances of an eclipse: its date and time, the geographical locus of observation, and the duration of totality or partiality. This is not the place to go into such detail, but the interested reader may refer to the appendix at the end of the book. It explains why the eclipse of 30 June 1973 was so exceptional.

## "The Sun Was Put to Shame and Went down in the Daytime"

In the ancient city of Ugarit on the Syrian side of the Mediterranean, archeological excavation turned up a clay tablet telling of an eclipse of the Sun, perhaps the most ancient ever to be reported.[3] Today, by careful calculation, we can reconstruct the exact date: it was 3 May 1375 BC. We read: "On the day of the new moon, in the month of Hiyar, the Sun was put to shame, and went down in the daytime, with Mars in attendance." Indeed, the brightest stars and planets become visible in the darkness that prevails during totality. Less than a century later, five solar eclipses occurring between 1226 and 1161 BC were reported in China on oracle bones found in excavations at Anyang (Henan Province).[4] Today, the term *ri shi* (日食) is Chinese for 'eclipse', but it also means 'lost Sun', a reference to ancient myths of dragons devouring the Sun during the eclipse. Drums were beaten, while archers

[3] It may be that the earlier eclipse of 21 October 3784 in India was reported in a chronicle. In P. Guillermier and S. Koutchmy, *Eclipses totales: histoire, découvertes, observations*, p. 97. See the bibliography.

[4] http://history.cultural-china.com/en/183History5571.html

shot arrows toward the Sun to scare the dragon away. And beware the astronomer of the imperial court if he made any mistake over his predictions, for his head would roll. In those days, it could be a dangerous job.

Ulysses may also have observed a total eclipse near the Greek island of Ithaca on 16 April 1178 BC, if we interpret in this way the mention of darkness at noon recounted in Book XX of the Odyssey[5]: "See too crowded with ghosts is the porch, ghosts hurrying down to the darkness of Erebus. Out of heaven, withered and gone is the sun, and a poisonous mist is arising."

But let us leave aside the slow scientific progress in the prediction of the dates and loci of eclipses, a long story that goes back to the Babylonians, the Greeks, the Indians, the Chinese, the Mayas, and the Arabs, passing through the world of the Renaissance and finally arriving at the astounding accuracy of modern computation. And it is thanks to the latter that ancient descriptions of eclipses, dated in the chronicles of the day on the one hand and in a universal chronology by celestial mechanics on the other, that we can situate the historical events of the distant past in a single chronology.

## An Astronomer at the Court of the Son of Heaven

Ancient science and the new science of the Renaissance, the Christian West and the Confucian East, came together in a quite remarkable way at the beginning of the seventeenth century in the context of a solar eclipse. Born in Macerata in central Italy in 1552, Matteo Ricci was a Jesuit priest who set off for India, then China, aboard a Portuguese ship. The only Chinese port then open to trade was the town of Macao, where the Portuguese were extremely active. Ricci learnt Chinese and dreamt of making his way to the heart of the Middle Kingdom, the northern capital of Peking or Beijing. Naturally, his aim was to introduce the Chinese people to the Gospel, but he quickly understood that he must first get a better understanding of this country and learn to love it. After a long period spent studying the Chinese culture, he finally arrived in Beijing in 1601 and soon made friends with the mandarins, the local intellectuals, sharing with them the fruits of European mathematics and astronomy, something he had studied at length before leaving.[6] Before long, Ricci became the tutor of the son of the emperor Wan Li, teaching him the science of the Western Renaissance. He produced a Chinese version of Euclid's *Elements*, the great classic of Greek geometry, and going the other way, a Latin version of the *Analects* (*Lun Yu*) by Confucius, then unknown in the West. From geographical knowledge based on accurate measurements, he drew up a map of the world as it was known at the time, in the form of a planisphere (Il Mappamondo,

[5] C. Baikouzis, M.O. Magnasco: Is an eclipse described in the Odyssey? (2008). http://www.ncbi.nlm.nih.gov/pubmed/18577587.

[6] Michel Masson: *Matteo Ricci, un jésuite en Chine*, Paris, Facultés jésuites de Paris éd. (2009).

Complete Map of the Earth's Mountains and Seas), which he offered to the Emperor and which was subsequently disseminated throughout Asia. He even dared to place the Middle Kingdom off centre.

Apart from his geographical knowledge and his mastery of standard instruments for measuring the positions and directions of stars, Ricci had tables called ephemerides, refined by the Portuguese for navigation purposes. These gave the motions of the Earth (diurnal rotation about its own axis and annual rotation about the Sun) and of the Moon (rotation about the Earth during the lunar month). And even before arriving in Beijing, he explained why the partial eclipse observed in Nanchang in the south of China fell short of what was predicted by the Emperor's own astronomers. In Ricci's own words[7]: "Many people came to ask me about this: was there an error in the calculation or had other causes reduced the degree of partiality? I took the opportunity to explain how an eclipse of the Sun is not universal, since it can be greater at one place than at another. Hence, it might happen to be greater where the calculation was carried out [in Peking], while being less in this much more central province [Jiangxi]. They were happy with this explanation, because it did not contradict the king's mathematicians, and because the reasons I gave them for eclipses of the Sun or the Moon seemed satisfactory to them. All this was unknown to them, especially the cause of the eclipse of the Moon, which their savants had never identified." Later, in Beijing, the solar eclipse of 11 May 1603 provided him with a fabulous opportunity for scientific demonstration. While the imperial astronomers had predicted the beginning of totality with an error of three quarters of an hour and the end with an error of a quarter of an hour, Ricci's prediction was significantly more accurate, and this earned him a tremendous reputation.[8] In 2010, exactly four centuries after the death of Matteo Ricci, I visited his tomb with some emotion, in the Jesuit cemetery, preserved today in the Party School of the Central Committee of the Communist Party of China in Beijing (*Beijing shiwei dangxiao*). During the Cultural Revolution (1966–1976), the magnificent stone monument which marks the tomb was buried underground to protect it. This Jesuit astronomer, whose remains were finally laid to rest in a China he had come to love so passionately, had lived three and a half centuries before us, and during this intervening period, eclipse calculations had never ceased to improve, right up to our rendez-vous above Africa on 30 June 1973 (Fig. 1.1).

---

[7] http://www.matteo-ricci.org/Opus/opus23.html

[8] Despite a considerable effort, the present author has been unable to find any record of the residual error in Ricci's prediction for the 1603 eclipse. However, the prediction by the Emperor's astronomers is reported with great accuracy in *Observations mathématiques, astronomiques, géographiques, chronologiques et physiques, tirées des anciens livres chinois ou faites nouvellement aux Indes, à la Chine et ailleurs, par les Pères de la Compagnie de Jésus. Rédigées et publiées par le P. Etienne Souciet*, Rollin, Paris (1732). Ricci's more accurate result cannot be doubted, however, as attested historically by the reputation he gained thereby.

**Fig. 1.1** The tomb of Matteo Ricci, preserved and honoured in the former Jesuit cemetery, located in the heart of Beijing in the Party School of the Central Committee of the Communist Party of China. © Pierre Léna

## A Cuckoo Flies Over Paris

It is 1912, at the end of the *Belle Epoque* and before the massacres of the Great War. Six years after the courageous attempts of Clément Ader (1897), the Wright brothers finally overcame the difficulties of flying a vehicle that was heavier than air. The French became passionately involved in this new science of aviation. The creators of sometimes strange flying machines exercised their imagination, and often their ingenuity. And then a 'diamond ring' solar eclipse was announced. It was to occur over the Paris area on 17 April. No further encouragement was needed to inspire two young

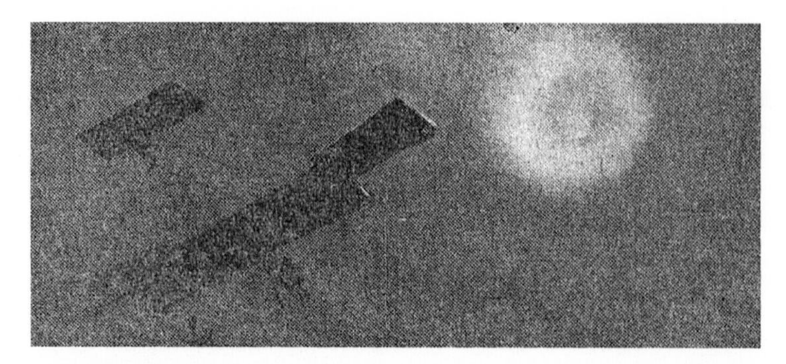

**Fig. 1.2** The first flight in history (1912) to observe a total eclipse of the Sun. The Voisin biplane was photographed from the ground in the direction of the eclipsed Sun. Taken from the daily newspaper *Le Matin*. Le Matin newspaper of 18 April 1912

men, Michel Mahieu, already an accomplished pilot at the age of 20, and Gaston de Manthé, his navigator, to take to the airs and view it from their military-style Voisin biplane, specially built for Mahieu. The plane took off from the landing strip at Issy-les-Moulineaux, crossed the Seine close to the Eiffel tower and, at an altitude of about 250 m somewhere near Saint-Germain-en-Laye, its two passengers encountered totality on the line of centrality during the short transit of the lunar shadow. Their flight lasted 50 min, so they were able to observe totality at 12 h 20 and land 10 min later at Issy-les-Moulineaux. The next day, the newspaper *Le Matin* gave homage to these two men "enchanted by having been able to observe the eclipse from a little closer than ordinary mortals", and above a light mist. The 18 April issue of *Le Matin* published close to a million copies containing an exceptional photo taken by an anonymous photographer. In it, we see the silhouette of Mahieu's Voisin biplane next to the eclipsed Sun. Despite the poor quality of the reproduction, we may surmise that it was the diamond necklace that caught the photographer's eye through the mist. This ring of bright beads around the Sun is observed when the apparent size of the lunar disk is exactly such that the Sun's light can just reach us through the valleys between the lunar mountains standing out from the rim of the Moon's disk (Fig. 1.2).

When the Great War broke out, Michel Mahieu gave his plane to the nation and then left to fight. As a captain, awarded the *Croix de guerre*, he was mortally wounded in the Somme at the age of only 26 years. An impressive statue of him and his brother August was put up in his home town of Armentières. Although he had no scientific pretensions, Michel Mahieu nevertheless made the first airborne observation of an eclipse, and he did it in France. Mahieu's Voisin had a wingspan of 15.75 m, and Concorde, in which we were to fly 60 years later, less than twice that (25.60 m). So Mahieu was our precursor, and it seems right therefore to tell his story here, as brought together by Robert Morris, a Canadian teacher and engineer himself fascinated by eclipses and aircraft![9] (Fig. 1.3).

[9] L. Robert Morris: *Biplan Voisin: l'éclipse de 1912, la naissance de l'astronomie en avion*. See the bibliography.

**Fig. 1.3** View by Canadian artist Don Connolly of the Voisin plane flown by Michel Mahieu as it crossed the Seine to join the line of centrality of the eclipse over Saint-Germain-en-Laye. *Inset:* Parisians watching the eclipse. © Don Connolly, Sydenham, Canada. Painting The Birth of Aviation Astronomy (Paris, 17 April 1912). Conception by L. Robert Morris. Composition by Don Connolly and L. Robert Morris. Acrylic on board (61 × 46 cm), 2012

Seven years later, during the total eclipse of 29 May 1919 observed in Principe off the coast of Gabon, 2000 km south of the path that Concorde would follow half a century later, the observations were used to check the calculations of Albert Einstein, based on his general theory of relativity published in 1916. Throughout the whole of history, this was without doubt the most fruitful eclipse for the progress of science.

## The Mystery of the Solar Halo

From the nineteenth century, eclipses began to interest astronomers for another reason. Up until then, the aim had been to make ever more accurate predictions of the time and place, thereby putting to the test calculations of the lunar and terrestrial motions. But now it was the inner workings of the Sun itself that eclipses could throw some light upon. The Greeks had already noted the presence of a weak halo of light which became visible around the eclipsed star. Describing the total eclipse of 98 AD, the historian Plutarch wrote: "In eclipses [of the Sun], the Moon allows a part of the Sun to spill over, and this reduces the darkness." This 'part', known today as the solar corona, was of great interest to astronomers, but it would long remain mysterious. It was thought that it might be the atmosphere of the Moon, or smoke, and it was only during the nineteenth century that this weak light would be correctly attributed to the atmosphere of the Sun itself, whereupon it became known as the solar corona.

## Tourism for Amateurs and Science Above the Clouds

Following the success of Michel Mahieu's flight in 1912 and the ensuing rapid development of aviation, the advantages of observing total eclipses from craft that could fly ever higher soon became a major motivation, either simply to experience the intense emotions procured by such a spectacle, or to penetrate the mystery of this fleeting solar corona.

On 10 September 1923, in California, the US Navy flew 16 planes, including one seaplane, to determine as accurately as possible the central axis of the path followed by the lunar shadow during a total eclipse of the Sun. One of the pilots was a certain Albert W. Stevens. In 1937, at which time he already held the world altitude record in a balloon, he took a Douglas DC-2 up to almost 9000 m above the Peruvian Andes, photographing for the first time this solar corona whose origins were still so enigmatic. These airborne observations were taken further after World War II, in 1945 and during the following years. On 30 June 1954, two saros[10] periods before the eclipse to be observed from Concorde, the British astronomer D. E. Blackwell set to work at an altitude of 9000 m aboard a propeller-driven Lincoln bomber. Indeed, he observed the sky with the door of the aircraft open and in the greatest discomfort, close to passing out, but without the porthole getting in the way of the light from the corona. The following year, the French astronomer Raymond Michard successfully made observations of the solar corona during the eclipse of 20 June 1955 while flying aboard a refitted Nord 2501 military transport aircraft in Indochina. For the eclipse of 20 July 1963 in the Canadian north, NASA flew a Douglas DC-8 powered by four jet turbine engines at close to 1000 km an hour, while up to then slower, propeller-powered aircraft had only been able to gain a few seconds by pursuing the Moon's shadow. In 1965, the same NAA team fitted out a Convair-990 plane, Galileo I, for astronomy and eclipse observation, while the US Air Force did several eclipse expeditions from 1965 on (Fig. 1.4). We shall return to these planes later.

The advantages of these airborne solar observations are clear enough. Any point on Earth can be reached by plane, wherever totality happens to occur. At higher altitudes, the probability of cloud cover is significantly reduced, the Earth's atmosphere is much more transparent to celestial light, and the sky is darker, whence the

[10] A saros period is 18 years, a characteristic interval between solar eclipses. See the appendix at the end of the book.

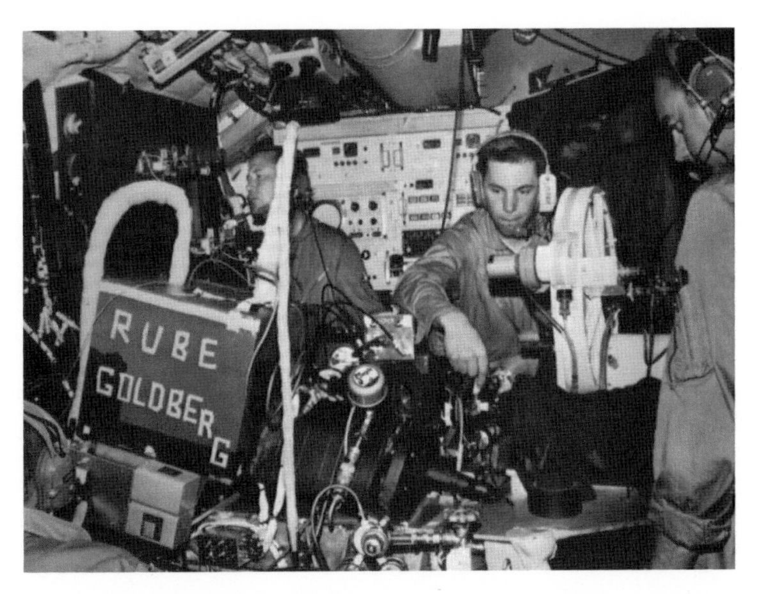


**Fig. 1.4** Above Samoa in the Pacific in 1965, observation of a total eclipse of the Sun by the team from the Los Alamos Laboratory in New Mexico. The scientific equipment is set up in the cabin of a four-jet Boeing NC-135 of the US Air Force, the Sun being viewed toward the left through a specially made porthole. In 1973, Donald Liebenberg (*centre*) was one of the foreign scientists invited to take part in the Concorde flight. © Los Alamos Scientific Laboratory/Don Liebenberg

corona stands out better. Finally, the duration of the eclipse can be slightly increased thanks to the speed of the aircraft as it follows the shadow. But there are disadvantages too, since the light must in principle pass through a porthole which seals the inside of the plane off from outside, while vibrations and the motion of the aircraft can disturb the stability of the onboard telescope used to provide an image of the Sun.

These first flights, often spectacular, led to an improved although still rudimentary understanding of the solar corona and its inherent physical mechanisms. They would thus supplement the many expeditions to observe eclipses from the totality region on the Earth's surface.

# Chapter 2
# Concorde: A Dream Takes Off

While astronomers sought enthusiastically to unravel the secrets of the Sun, the stars, and the universe in general, other men and women dreamt of building machines that could tear themselves away from the Earth's surface and take to the skies, going ever higher, ever faster, and ever further. These were the heros of the twentieth century, inspiring us even before the first men went into space (Yuri Gagarin in 1961), then to the Moon (Neil Armstrong and his colleagues in 1969), leaving the cradle of humanity to explore new, still more chimerical horizons.[1] The supersonic plane Concorde was part of this saga of the last century, in which the ambition of intercontinental travel for a hundred passengers, putting New York at only three and half hours from Paris, would finally be realised. And it was precisely in this year of 1969 that the Concorde prototype 001 made its first test flight from Toulouse airport in Blagnac, 2 months after its Soviet rival, the Tupolev Tu-144, took to the air for 38 min near Moscow.

## Entente Cordiale and a Challenge

The decision to build a supersonic passenger plane was not an easy one to reach. There had been two jet passenger planes since World War II, exploiting the technological progress stimulated by this battle of giants. The first was the British four-jet Comet, which took off in 1949 and was in service from 1952. Two tragic accidents marred its commercial career, leaving the way open for Boeing's long-haul four-jet 707. In Canada, a four-jet passenger plane, the Avro C102, a contemporary of the Comet, was also developed, but it was abandoned in 1951 in favour of a supersonic fighter plane. Other four-jet passenger aircraft would come later: the Vickers VC-10 in the United Kingdom and the Ilyoushin IL-62 and IL-86 in the

[1] Roger M. Bonnet: *Les Horizons chimériques*, Dunod, Paris (1992).



P. Léna, *Racing the Moon's Shadow with Concorde 001*, Astronomers' Universe,
DOI 10.1007/978-3-319-21729-1_2

Soviet Union. At the same time, the French Caravelle, a twin-jet aircraft with a quite different design, was extremely successful. Its first test flight was in 1955 and it was used by Air France from 1959. Later in this book we shall meet a Caravelle in the service of astronomy.

At the end of the 1950s, military aircraft were just beginning to explore the supersonic range, exceeding the speed of sound, some 1000 km an hour, in the atmosphere, although needless to say, breaking the sound barrier raises serious problems for the aerodynamics of an aircraft. At the height of the Cold War, the US military had a B-58 bomber that could fly at Mach 2, that is, twice the speed of sound, while carrying an atomic bomb. When this supersonic four-jet aircraft travelled non-stop from Fort Worth in Texas to Le Bourget in France in 1961 in just 3 hours and 19 min, Lindbergh's maiden transatlantic flight was much in mind. But although it flew at Mach 2 for some of the time, the aircraft had to be refuelled in flight twice over the Atlantic. A week later, this plane crashed at Le Bourget during a flight demonstration and its three crew members were killed. There were many losses among the hundred and twenty-eight B-58 aircraft and their crews, since 2 years later only 95 of those planes were still flying.[2] Already engaged in the race to the Moon, the United States also wanted to take its place in supersonic passenger transport, but it failed, and the project of a Boeing 2707 (SST) was abandoned in 1971. In France as in the United Kingdom, the design offices were also pursuing the possibility of a high speed commercial aircraft. In France, a four-jet project called Super Caravelle was begun in 1961 and a model was presented at the Paris air shown in Le Bourget. The plane would have carried a hundred passengers over distances up to 4500 km, still not enough to get from Paris to New York, at a speed of Mach 2.2, that is, more than twice the speed of sound. As one can easily imagine, transporting bombs at Mach 2 with a potentially disposable military crew bears no relation to the reliability required for a commercial airliner carrying men, women, and children.

General de Gaulle, whose ambitions for France were well known, realised that France alone could not cover the whole cost of developing such a plane. In November 1962, a few weeks after refusing to allow the United Kingdom to enter the European Common Market, France signed an agreement with its British partners to build a supersonic aircraft for commercial use.[3] This would become Concorde, and not Concord, after long and arduous negotiations about whether to use the English spelling with no 'e' at the end. In France, the company Aérospatiale was made responsible for the programme, while André Turcat, pilot and engineer from the *Ecole polytechnique*, was chosen to direct flight tests. In 1959, Turcat had beaten the world speed record over a closed loop of 100 km in a ramjet-powered

---

[2] Kenneth Owen: *Concorde: New Shape in the Sky*, Janes, London (1982).

[3] The birth of the Concorde project and the negotiations between the two countries are described in the book *Concorde [ICBH Witness Seminar]*. See the bibliography. The book contains first hand accounts by André Turcat and Henri Ziegler.

**Fig. 2.1** Image of Concorde's wings produced by detailed aerodynamic studies, using either computation or models, at the *Office national d'études et de recherches aérospatiales* (ONERA). A model in landing configuration is placed in a flow. Coloured liquids are emitted at different points. Their trajectories reveal the interactions between the plane and its fluid environment, e.g., the production of turbulence. © ONERA The French Aerospace Lab

Griffon II aircraft.[4] We will meet up with André Turcat for the solar eclipse 15 years later (Fig. 2.1).

## Concorde Takes Off

There were two prototypes of Concorde, one in France, the other in the United Kingdom, thus sharing the work of exploring the many technical issues that needed to be solved: materials, wings, engines, landing gear, safety systems, flight procedures, and so on. And this was no small challenge. The military aircraft then in service and able to break the sound barrier fell far short of the range, the payload, and the long duration of supersonic flight assigned to the project. The new plane would have to cruise in normal engine regime without the afterburn[5] used by military aircraft for a few minutes to increase the thrust of the jet.

[4] The ramjet is the simplest conceivable jet engine. The gases produced by the burning fuel are simply ejected at the outlet of a tube and this produces the thrust. No moving parts are required, but this jet engine cannot be used for take-off. http://en.wikipedia.org/wiki/Ramjet.

[5] Afterburn or reheat is a fuel-costly method for increasing the thrust of the jet engines for a very brief time by injecting fuel directly into the exhaust gases, where they are set alight. See the appendix on Concorde at the end of the book.

André Turcat and others have told the story of the preliminary work. On 2 March 1969, under the eyes of General de Gaulle who was watching on television, the prototype, a beautiful white bird, took off from Toulouse at a speed of 280 km/h (150 knots) with André Turcat and his copilot at the controls, while the flight test engineer Henri Perrier and the flight test mechanic Michel Rétif watched over the four engines and the 136 tonnes of machinery. Turcat writes: "For us, it was as though we barely spent a moment in introspection, for all our senses were straining over the instruments ... There was no room for apprehension, only action; to the point where I wouldn't even experience the usual primitive pleasure of take-off, with all the huge power held in one hand operating four throttles."[6] We will meet up with three of these men 4 years later during the eclipse. The day after this first successful flight by Concorde, NASA astronauts sent up a rocket for the Apollo IX flight, during which they would test all the equipment of the lunar landing module in orbit around the Earth prior to the successful lunar mission of Apollo XI. Thus, in his brief and modest speech after landing before an enthusiastic crowd in Toulouse, André Turcat sent them warm congratulations. On either side of the Atlantic, two of the most unlikely adventures of the twentieth century were on their way to the sky. One would allow a hundred passengers to cross the ocean every day in just 3 h and 30 min, while the other would take three men to the Moon in about a hundred hours.

On 1 October, Concorde cautiously crossed the sound barrier (Mach 1) without mishap, and the next day, at the Elysée Palace, the new President of the Republic Georges Pompidou confirmed to the president of Aérospatiale Henri Ziegler, accompanied by André Turcat, his determination to pursue the Concorde programme. The flight tests went on and the following month, during flight 102 above the Atlantic, Concorde 001 finally reached Mach 2 for a period of 53 min. This was a speed still largely unknown for such a heavy aircraft, apart from the unfortunate B-58 of the US Air Force.

André Turcat, always attentive to detail, had been able to analyse the fatal accident of the B-58 in Le Bourget, and had drawn invaluable conclusions for piloting his own test flights. The formidable gamble of the Concorde project had been pulled off, a scientific and technical gamble, also requiring considerable courage on the part of the test flight crews and visionary boldness by industrial stakeholders. But the commercial challenge was yet to be met.

Doubtless this victory, both technical and political, owes much to the close relationship between the French people and the world of aviation. In the 1960s, who had forgotten Clément Ader's first motorised flight in Issy-les-Moulineaux in 1897, even though it had been contested? Or indeed, the successive records achieved in the 1900s, and Blériot's channel crossing; or the flying aces of the Great War, such as Guynemer; or the Aérospatiale epic, connecting Toulouse to Santiago in Chile; or Nungesser and Coli who disappeared in 1927 while trying to cross the Atlantic from east to west in their *Oiseau blanc* (White Bird); or Saint-Exupéry, pilot and writer? This public awareness of a historical tradition was as much felt by the

[6] A. Turcat: *Concorde: essais et batailles*, p. 143. See the bibliography.

powers that be as by industry and its most modest worker, and it undeniably underpins so many of these programmes that the French people have reason to be proud of, such as Caravelle, Concorde, and Airbus, to name but the civil aircraft.

From then on and until the scientific flights of 1973, Concorde 001 would continue to undergo tests to define the production aircraft, the first of which took off at the beginning of 1975. But 001 also hosted VIPs, went from Paris to Dakar in 2 h and 35 min, and did a trip to Brazil and Argentina. At the beginning of 1972, the prototype 001 was being prepared for retirement, but the people of Toulouse were by then used to seeing the pure white delta (or more precisely, 'gothic') wing and the long pointed droop nose, like the beak of a strange bird, lowered to give the pilot the visibility required for landing. This was the moment when science and astronomy would enter the lives of the Concorde 001 and its crew in a quite unexpected way.

## In Which a Tragedy Is Endured, but the Business Imperative Wins Out

By 1976, the American and Soviet supersonic projects had been abandoned, whereas the commercial career of Concorde began in January, with a dozen aircraft flown by Air France and British Airways. This part of the story was rather different, unfortunately marred by the United States' refusal to allow access to JFK airport in New York until 1977. There were no accidents during this period, while tens of thousands of admiring passengers were transported, up until the tragic take-off of 25 July 2000 at Charles-de-Gaulle airport in Paris. A tyre burst on the Air France F-BTSC carrying a hundred passengers and nine crew members, and a fragment of it hit the wing, causing a kerosene leak, a fire, and the loss of two engines. The plane crashed into a hotel in Gonesse 2 min after take-off, adding four more victims on the ground. While driving my car in the streets of Paris, I heard the radio announce the crash and my heart sank as my mind was suddenly filled with images of the eclipse flight almost 30 years earlier. All Concordes were immediately grounded during the ensuing enquiry, but they regained their airworthiness certificate in August 2001. However, commercial exploitation was halted indefinitely by both Air France and British Airways in 2003. The economic health of the world was no longer what it had been in the 30 prosperous years[7] following the end of World War II. And so ended the dream of a worldwide fleet of more than 1500 supersonic airliners flying in 1990, as suggested in a book[8] published in 1971 and prefaced by Didier Daurat, the founder of the legendary 1930s line to Santiago, Chile, represented by the

[7] Known as the *Trentes Glorieuses* in France.

[8] In *La Grande aventure de 'Concorde'*, Presses Pocket, Paris (1971), a book prefaced by Didier Daurat, founder of Aérospatiale, the author Jean-Pierre Manel announced the prospect of a world supersonic fleet of some 1600 supersonic planes by 1990.

character Rivière in Saint-Exupery's *Night Flight*. Now passengers from the world over who pass through Roissy Charles-de-Gaulle airport can see Air France's F-BVFF Concorde exposed nose up, as if ready for take-off.

Note: The reader will find several technical details about Concorde in appendix at the end of the book.

# Chapter 3
# This Dark Brightness that Falls from the Stars

## In Which a Navy Officer Becomes an Astronomer

Jack Eddy was born in the country of the Pawnee Indians, in a small town in Nebraska (USA), in 1931, of a modest family who would be hard put to finance the studies of their three children with the income they earned from the local farmers' cooperative, supplemented by his mother's salary as a primary school teacher. But Jack nevertheless gained a place in the US Naval Academy and would soon become an officer, particularly interested in what he would have to know from astronomy in order to take a sight at sea, and much more. Stationed aboard an aircraft carrier, he took part in the Korean war, but it was one night in the Atlantic, while on surveillance duty on a destroyer, that his life was suddenly turned upside-down. A man had just fallen overboard and Jack requested the order to go back and rescue him from the water. But his request was turned down, the ship continued on its way, and the drowning sailor disappeared into the dark night. Immediately, Jack the pacifist, so handsome in his officer's white uniform, decided to leave the Navy and devote his life to science, and in particular, to astronomy. Discharged from his military duties, he entered the university of Colorado, and in 1961 prepared a doctoral thesis under the supervision of an astronomer who happened to be one of the leading experts on that part of the Sun which 'spills over' during an eclipse, as Plutarch had noted so long before.

I met Jack and his wife Marjorie a few years later in the observatory where I first came into contact with the world of aviation, at the foot of the Rocky mountains, dominating the Great Plains where the Pawnees and their bisons have unfortunately been replaced by motorways and cheap motels. It was Jack, whom we shall encounter again later, who told me his story, and another one too, that of Thomas Edison.



P. Léna, *Racing the Moon's Shadow with Concorde 001*, Astronomers' Universe,
DOI 10.1007/978-3-319-21729-1_3

## In Which a Brilliant Inventor Is Engulfed in Feathers During an Eclipse

Born in the middle of the nineteenth century in Ohio, Thomas Edison, fascinated by experimental science since he was a child, would become a brilliant inventor and businessman.[1] Humankind owes him the first phonograph, and also the first electric light, with a filament which becomes incandescent when a current passes through it, transformed into a source of white light. He filed more than a thousand patents and founded the industrial empire that went by the name of General Electric. At 22h00 on the evening of his funeral—he died at the age of 84–, President Hoover had all the lights in America turned off in homage. But Thomas Edison also had a brief encounter with astronomy at the age of 31, when the astronomer Samuel Pierpont Langley offered a challenge to the young but already famous inventor, warning him, however, that there would be no financial reward, only that he would do a service to science. Edison would have to invent a highly sensitive instrument to detect, during the next total eclipse of the Sun, the weak infrared light thought to be emitted by the solar halo—the corona. Invisible to our eyes, although some snakes do detect it with theirs, infrared light is the radiation that stimulates in us the sensation of heat when we expose our skin to the Sun, or to a wood fire. This infrared light, which we shall discuss further below, was largely unknown in those days.

So Edison set to work and invented the tasimeter, a strange instrument in which the tiny amount of energy received in the form of infrared light would cause the expansion of a piece of rubber by heating it. This in turn would press upon a piece of graphite, whose electrical resistance would thereby change in a way that could be measured. Although he was not a professional astronomer and had little experience in observing the stars, the young Edison, already a star himself in the eyes of the press and public opinion, decided to organise his own eclipse expedition during the summer of 1878, sure that his experimental genius would do much better than all those 'mathematicians'. He joined a group of genuine astronomers in Rawlins, Wyoming, a small town on the railway line on its way to the Far West, to observe the solar halo for himself. Having set up his instrument by the door of a small chicken hut to protect it from the wind sweeping across the plain, he carried out a successful night time test on the star Arcturus. Then, in the afternoon of 21 July 1878, he pointed it at the solar halo for the 3 brief min of totality. And lo and behold! The tasimeter did indeed pick up a signal from the Sun (Fig. 3.1).[2]

The rest of the story involves some controversy. The first issue, anecdotal but raised by Edison himself, although he failed to give any details, proved extremely popular with the press over the following few days: Edison claimed to have forgotten that the darkness of a total eclipse disturbs animals, and that the chickens,

[1] J. A. Eddy: Edison the scientist. In: Applied Optics **22**, 3737 (1979).

[2] J. A. Eddy: Thomas A. Edison and infrared astronomy. In: Jour. Hist. Astr. 3, 165 (1972).

**Fig. 3.1** Edison's tasimeter. A horn picks up the infrared light coming from the *left*. It falls on a vertical bar and heats it up, thereby squeezing a piece of graphite whose changing resistance can be measured. Courtesy of the American Journal of Science (1879), cited by J.A. Eddy, see note 19

feeling that night was coming on, had rushed back to the chicken hut in a flurry of feathers, knocking into the crucial instrument and its horrified owner as they went by. The second controversial issue was a question of priority, as is often encountered among scientists, in proportion to their feelings of self-importance. The young Edison, although inexperienced as an astronomer, nevertheless claimed to be the first to measure the faint infrared light from this solar halo, thanks to the extreme sensitivity of his tasimeter. This claim, taken up by the press, was immediately contested. An astronomer from Milan, professor Luigi Magrini, had already found evidence for this radiation[3] during the brief total eclipse of the Sun on 8 July 1842, visible in the south-east of Europe 3 years prior to the first photograph (daguerreotype) of the Sun. But Edison had other more important business than digging around in observatory libraries to discover a predecessor who might have overshadowed him! (Fig. 3.2).

Whether or not it was the first time in the history of astronomy that an eclipse had been observed in the infrared, this expedition to the Great Plains of the American west already contains all the ingredients that will occupy us throughout the rest of our story: infrared light, the mysterious solar halo, and above all, the eclipse itself. As far as aircraft are concerned, we have already spent some time on them, and yet here they return.

[3] http://sites.google.com/site/histoireobsparis/table-des-matieres/chapitre-7-arago-directeur-des-observations/eclipse-et-premier-daguerreotype-du-soleil

**Fig. 3.2** Numbered points are those where the coronal infrared emission was measured during the eclipse. Their positions are shown against a photograph of the solar corona taken during this same total eclipse of 29 July 1878 in Wyoming (USA). Courtesy of the US Naval Observatory (1880), in American Journal of Science, op.cit.

## Going to the Stratosphere to Look for Stars

If I was in this observatory in the Rocky mountains in 1968, where I became acquainted with Jack the former navy officer, while every evening on the television I could watch in horror the shelling of Vietnam by American bombs with their defoliants, it was because of my own interest as a young astronomer in infrared light.

Right at the beginning of the nineteenth century, in 1800, the British astronomer of German origin William Herschel had produced a rainbow by dispersing the white light of the Sun in a prism, then slowly moved a thermometer from one side of this rainbow to the other, from violet and indigo across to red. For each colour, the rise in temperature of the thermometer indicated the energy absorbed from the Sun in that colour of light. To his great surprise, Herschel observed that the thermometer reading continued to rise when he moved it beyond the red, thus indicating the presence of radiation that the eye could not perceive. He had just discovered the existence of infrared light. The distribution of this light, known as its spectrum, extended over a very broad range which would long remain unexplored, from visible red light to radio waves. In order to reveal the infrared radiation of remote bodies, it was essential to use much more sensitive thermometers than Herschel, and also to have large telescopes. Even in the middle of the twentieth century, when I was taking my first steps as an astrophysicist, the only infrared light yet detected from celestial sources was from the Sun, the illuminated and hence hot face of the

Moon, or indeed the planets Venus and Mars. To complicate matters, even when there is no cloud cover, the water vapour present in our atmosphere only lets through a very little of this kind of light. The only solution was to observe from the tops of high mountains, or from very dry desert regions, so collecting this light without too much loss. Better still, by sending a balloon up to an altitude of 20 or 30 km into the stratosphere, or observing from an aircraft at a height of 10 km, one could considerably increase the chances of a favourable observation, although of course there was a price to pay in terms of the technical complications as compared with an observatory firmly fixed to a mountain top.

In California in 1968, aboard NASA's powerful four-jet aircraft Galileo I, dedicated to other aspects of astronomy as well as eclipse observation, I learnt to fly high to steal some snippets of infrared light from the Sun and stars. On the night before the flight, while making the final adjustments in the desert hangar where the plane was kept, standing close to the long silver fuselage which held the instruments that Jack had helped me to design and breathing in the strong smell of oil, I had a marvellous feeling of stepping into unexplored territory. I was terrified at the thought of the failure that would result from the slightest error in the alignment of my telescope or the adjustment of our thermometers—in fact, thermometers have become utterly different compared with those of Herschel's day, now cooled to a temperature of 269° below zero to make them more sensitive, but let us not be concerned by that here.

Five years later, it was with great sadness that I learnt on 12 April 1973, while in Toulouse preparing the Concorde flight, that Galileo I had been destroyed on its return from a science flight with 12 men aboard, after a terrible collision with a small Navy aircraft, just before coming down on the landing strip where we had spent so many nights. Its successor Galileo II had a less tragic fate, without victims, a few years later, catching fire on landing after a science flight. Science is not always plain sailing.

It was 1969 and France was still feeling the consequences of the civil unrest of May 1968. The Paris Observatory and the Sorbonne, which I had just entered, were as shaken up as the rest of France, or even more so, with astronomers dreaming of putting the world to rights, if not the whole universe. So why not go on exploring the infrared sky with a plane? It was to be a Caravelle reserved for test flights, carrying the number 116, a marvellous twin-jet already mentioned in this story, that would host our telescope, specially built to fly on this plane, year after year, as high as possible, and at so many different latitudes (Fig. 3.3). Like Fabien in *Night Flight*, when he emerges into a star field[4]: "Their pale magnets attracted him. He had struggled so long for a glimpse of light that now he would not have let even the faintest get away from him. [...] And thus he rose towards these fields of light. [...] He wandered amid stars gathered with the density of a treasure, in a world where nothing else, absolutely nothing else other than he, Fabien, and his comrade, was living".

---

[4] Antoine de Saint-Exupéry: *Night Flight*, Chap. XVI.

**Fig. 3.3** From 1974 onward, the Paris Observatory flew a telescope aboard the Caravelle 116 stationed at the Brétigny test flight centre in Essonne, France, photographed here as it approaches the runway. The telescope, designed to observe stellar infrared radiation, is set up in the position of the port side emergency exit (Fig. 3.4). This Caravelle is now exposed at the aerodrome in Montélimar. © Centre d'Essais en Vol (Test Flight Center)

A smiling young man, Claude Nicollier, barely 30 years old, passionately interested in astronomy and aviation, had just joined our group from Switzerland.[5] Captain in the Swiss air force, he roamed the sky and the mountains at the speed of sound. After his doctorate, he became an airline pilot for Swissair, but he could not resist the attraction of astronomy and soon returned to it. Europe then selected him as an astronaut aboard the Space Shuttle. Our telescope and its observation programme were chosen to carry out an in-flight training programme for astronauts, prepared on the Caravelle and carried out aboard NASA's Galileo II plane in the USA (Fig. 3.4). With exceptional kindness, Claude taught me the patience and absolute rigour required for his job, qualities I seriously lack. A decade later, he went on three extraordinary astronomy missions into space aboard the Space Shuttle, including one to repair the Hubble Space Telescope.

In our work on Galileo I and II and on the Caravelle, we built up a strong and friendly team of researchers, students, technicians, and engineers (Fig. 3.5).[6] We

[5] http://fr.wikipedia.org/wiki/Claude_Nicollier

[6] D. Rouan: *Osiris et Astroplane: l'astronomie infrarouge aéroportée en France*. In: L'Astronomie no. 62, June 2012.

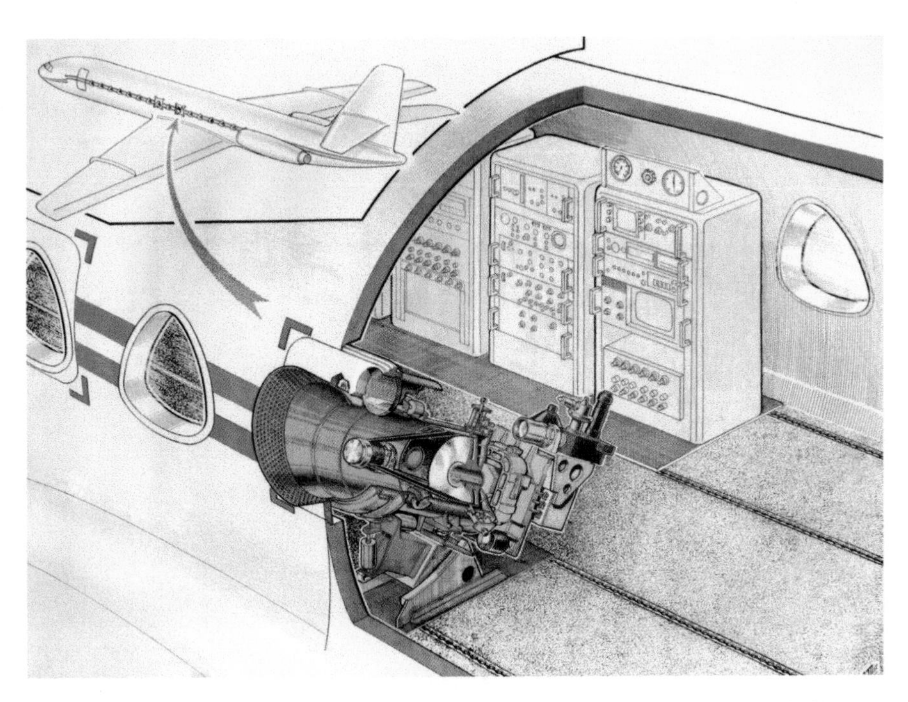

**Fig. 3.4** Artist's view of the telescope set up in the Caravelle 116 aircraft by the Paris Observatory, before being transferred to NASA's Galileo plane. Between 1972 and 1979, with my PhD students Daniel Rouan, Marie de Muizon and others, this instrument would make many astronomical observations of infrared radiation in both the northern and southern hemispheres. Drawing of J. Mir

had learnt how to fly toward the stars, our financial backers trusted us, and the infrared sky was opening up to our efforts and those of other astronomers around the world, with several discoveries on the horizon. So enter Concord 001.

**Fig. 3.5** The whole team from the Paris Observatory aboard NASA's Galileo I aircraft in 1976, with two of their European partners, gathered around the telescope (below the blue tube in the background). Many flights were carried out over California to explore the infrared sky. © P. Léna

# Chapter 4
# The Run up to the Longest Total Eclipse in History

The site in Blagnac near Toulouse is part of an aviation legend. It was here that Didier Daurat and Pierre-Georges Latécoère created the Aéropostale line to South America, made so famous by Henri Guillaumet, Antoine de Saint-Exupéry, and Jean Mermoz. It was from here that they took off from a short runway, now almost covered with buildings, with their precious cargo of mail, toward the south and the razzias of the African coast. And in Dakar, it was with great emotion that I came to contemplate the stone monument[1] to Mermoz and his companions who fell from the sky, now standing at the site of the landing strip where they came down before setting off across the South Atlantic in their Laté-28 seaplane, at the intersection of the *rue de la Pyrotechnie* and *avenue Cheikh Anta Diop*. And in this year of 1972, it was also from this site in Toulouse that the prototype 001 of Concorde would take off, week after week, up until the time a few months later when all the test flights had been successfully completed under André Turcat's leadership.

## A Crucial Lunch Meeting

That same spring of 1972, I made a phone call from my laboratory in Meudon, requesting a meeting with André Turcat who was in charge of the test flights, although I did not mention my reasons or discuss them with anyone. The man was already of considerable renown and highly esteemed, but he kindly accepted the request from this young and unknown astronomer, and very courteously invited me to lunch with him at the restaurant La Fontaine at the airport when I disembarked from the Caravelle used for my Air Inter flight from Paris. Perhaps this was a good omen? On the paper napkin covering the table where we lunched looking out over the landing strip, I quickly outlined my dream, sketching a map of West Africa and

[1] http://www.aerosteles.net/fiche.php?code-dakar-sacrecoeur-mermoz



P. Léna, *Racing the Moon's Shadow with Concorde 001*, Astronomers' Universe,
DOI 10.1007/978-3-319-21729-1_4

explaining that the Moon's shadow would be moving at close to the maximum speeds that could be reached by Concorde. I also detailed my modest experience of aeronautics and concluded by describing the extraordinary record which lay within our reach: the Sun would remain totally hidden to us for an incredible 80 min, more than ten times the longest time that would ever be possible from a fixed observation point on the Earth's surface. André Turcat gives the following account of this first meeting[2]:

> A relatively well ordered shock of black hair, with likewise black and shining eyes behind his spectacles, this astronomer had just sat down at the table before me and began to present his project at some length. 'All that was needed' was a supersonic aircraft, admittedly a little expensive, with a few holes made in the roof of the cabin, a few portholes made out of a special material, an electricity supply for his astronomical observation equipment, and extremely accurately programmed flights over Mauritania and Chad. Concerning the storms over the intertropical convergence zone, this young astronomer had little time for them. His final summary was brief: 'So what about it?' Yes, indeed! I was won over and my enthusiasm matched in every way that of this young professor. But my mind was already occupied with all the problems we would have to face.

When the young astronomer flew back to Paris that evening, a plan of action had already been sketched out, for the days were counted. It was May and the eclipse would occur in June of the following year. My laboratory was very busy with many other projects and I wondered how this one would be received by my colleagues. Who would finance it? Where would I find the much needed support and cooperation? When I got home, I was suddenly terrified by my own audacity and the risk of failure after having dared to involve such a prestigious aircraft. My wife encouraged me and I took the little globe I used to introduce our first two children to the fact that the Earth was round, trying to reassure myself by considering at length the possible trajectories (Fig. 4.1).

But before solving all these problems and resuming contact with the research unit in Toulouse that André Turcat had immediately put in the picture, there was one key question I had to answer. Mobilising Concorde and chasing after the shadow of the Moon was all very well, but to achieve what exactly? What scientific problem was I going to tackle? What unresolved question would I attempt to answer by this novel astronomical observation, made possible by the exceptional totality observed from the Earth's stratosphere?

[2] A. Turcat: *Un mythe éclipsé*. In: *Bulletin de l'Académie des sciences, agriculture, arts et belles lettres d'Aix-en-Provence* (2013). A less detailed account can be found in A. Turcat: *Concorde: essais et batailles*, pp. 244–248. See the bibliography.

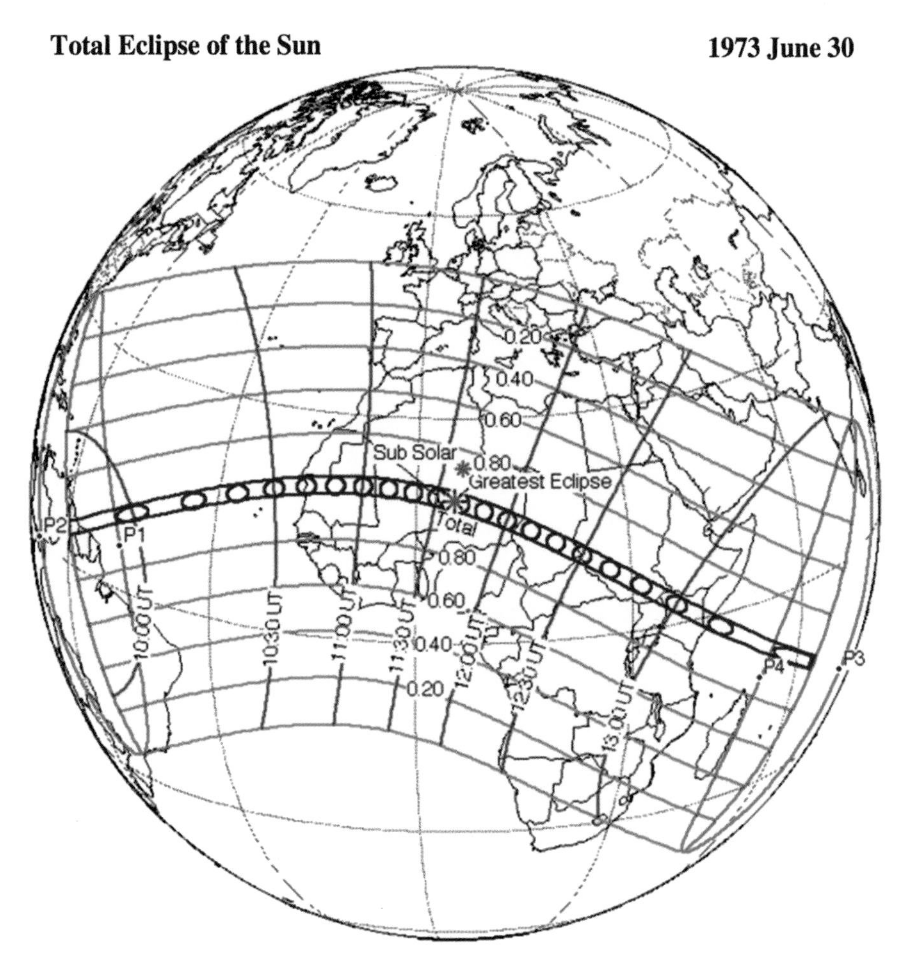


**Fig. 4.1** Dreams inspired by a small globe. The narrow path of the Moon's umbra from P1 to P4 (*dark blue*) and penumbra (*area covered by light blue lines*), as predicted by celestial mechanics. Credit: Eclipse Predictions by Fred Espenak, NASA's GSFC

## Corona, Dust, and Infrared Light

Interplanetary space, the space between ourselves and the Sun, is not completely empty of all matter. The Sun is constantly ejecting hydrogen gas, and there are already tiny dust grains in this space, fossils of a kind, attesting to the formation of the Solar System a few billion years ago, or residues left over from tails of comets when they come too close to the Sun and evaporate. It is these hydrogen atoms and dust grains that produce the solar halo that Plutarch had noted and my friend Jack had studied. Since it has become a subject of scientific investigation and detailed observation, this rather faint envelope has become known as the solar corona. Now at the time when the astronomy of infrared light was coming into being, dust, omnipresent in the universe, was beginning to arouse the interest of astrophysicists. Indeed, the telescope aboard the Caravelle had been set up to study it, and we had become specialised in the main techniques. While the existence of this interstellar dust had been suspected, or even established, for some time then, its low temperature meant that it was only observable through its infrared radiation, and I have already mentioned the difficulties involved in detecting this radiation with 'thermometers' that remained to be invented.

During an eclipse some time earlier (1966), an American colleague, Robert M. MacQueen, whom I had met while working with Jack Eddy on flight preparations for the four-jet Galileo I, had in fact been trying to repeat the observation of the solar corona carried out by Thomas Edison, and he had come up with a rather interesting result. Not far from the surface, the Sun seemed to be surrounded by rings of tiny dust grains—silicates, in fact—which concentrate in space before evaporating under the effects of the intense radiation from our star. These grains, measuring something like a thousandth of a millimetre, should emit a weak infrared light. They appeared to be concentrated in the plane of the orbits of all the planets, the so-called ecliptic, forming rings, or perhaps surrounding the Sun entirely in the form of shells. It was already understood without formal demonstration that, one way or another, over millions of years, these grains were capable of sticking together to form ever larger ones, and eventually exceptionally large ones like our own planet. They certainly deserved investigation in order to explain their presence, if confirmed, so close to the Sun. Would we thus be able to improve the result obtained by our colleague by exploiting the advantages of a flight that could provide more than a whole hour's worth of corona observations, rather than just a few brief minutes, without being dazzled by the intense radiation from the solar surface? Better still, flying at an altitude of 17,000 m in the stratosphere, Concorde would be well above the water vapour which is present in the lower Earth atmosphere and which significantly attenuates infrared transmission. We should be able to confirm the existence of rings and perhaps even determine their chemical composition. I discussed all this with my colleagues and made a few preliminary calculations, whereupon we agreed that it was feasible. Our subject would be the F-corona, that is, the dust in the corona. A little later, during the autumn, I went back to the United States and described the project to my colleague Donald Hall, a brilliant 28 year old Australian who had also specialised in solar infrared radiation. Enthusiastic, he accepted to join our project and got the agreement from the Kitt Peak Observatory in Arizona where he was working.

I now had a firm scientific motive for undertaking this adventure. It remained only to make a formal announcement, set out the details and feasibility, and find funding. And time was short, too short, for there were three irons to get in the fire: convincing the supervisory authorities, beginning the close collaboration with the research unit in Toulouse in order to determine a possible refurbishment of the plane, and building a completely new instrument suited to the kind of infrared observations we were hoping to make.

## In Which a Future Research Minister Gives His Support

I put together a small team thanks to the immediate support of close and trusted colleagues, thus reducing the very real risk of overload that my initiative might impose on the programmes already undertaken by the laboratory. From then on, Charles Darpentigny, Alain Soufflot, Jacqueline Mondellini, and Yves Viala were constantly by my side. Despite the extra workload, I continued to give my physics courses at the university, without daring to tell my students about this slightly crazy project that was in fact occupying all my thoughts. Naturally, the first thing I did was to get the agreement from the appropriate research authorities, without which I could do nothing. The president of the Paris Observatory was then Raymond Michard,[3] and as luck would have it, this eminent solar specialist, somewhat shaken by the events of May 1968, had also carried out an observation of a solar eclipse in Indochina 20 years earlier, aboard a Nord 2501 military aircraft. Needless to say, he gave us his full support, provided that it was not going to cost too much. The necessary financing could only come from higher up. Now, our laboratory also depended on the French national research organisation (*Centre national de la recherche scientifique* CNRS) and a completely new institute had just been set up in response to the upheavals of May 1968. Those in charge were astronomers, among them my friend Roger Bonnet, fully aware of both the interest and the risks of this slightly mad undertaking. But they gave their support.

I also went to see Pierre Aigrain, then called the Delegate General for Scientific and Technical Research, since there had never yet been a genuine minister for scientific research under the Fifth Republic in France. President Georges Pompidou had flown on Concorde 001 in 1971 to meet the then President of the United States Richard Nixon (invested in 1969) in the Azores. Regarding Pierre Aigrain, a future minister himself, he was the Prime Minister's acting Delegate General for this domain. He was a brilliant physicist, literally abounding in energy, who had taken me into his laboratory at the *Ecole normal supérieure* some 15 years earlier, helping me to develop a first infrared-sensitive 'thermometer' while I was still a student. I met him on a beautiful September morning and explained the project and the science we expected from it. Drawing as always on his pipe, it only took him a few minutes to pledge his support and some of the funds necessary for the construction of our instrument. The CNRS and the National centre for space studies (*Centre national d'études spatiales* CNES), the two large public bodies that financed our laboratory, also approved the underlying idea of the project, despite the vagueness pervading certain aspects of it. The Minister of Finance even agreed to exonerate the whole operation from the imposition of tax.

For his part, exercising his moral and technical authority, André Turcat took similar steps in Toulouse, approaching the managing director of Aérospatiale, Henri Ziegler. The latter, an ex-student of the *Ecole polytechnique* who had not forgotten his love for science, was Commander in Chief of the French Forces of the Interior (*Forces Françaises de l'Intérieur* FFI) in the French Resistance, and also a

[3] http://www.bureau-des-longitudes.fr/membres%20correspondants/michard.htm

former test pilot. At this time, he was also one of the founders of the Airbus programme. He viewed the eclipse project favourably, and later on, having received a full technical report, he accepted that Aérospatiale should cover the costs of the transformations required to refurbish the plane, as well as the cost of the eclipse flight over Africa and the necessary rehearsal flights.

So in autumn 1972, I was told that we could begin work, but that no firm decision about the eclipse flight would be taken before February 1973, when we would be in a position to make a final assessment of the preliminary studies and costs. So we only had 4 short months to prepare for the big day.

## How Many Holes Should Be Made in Concorde's Fuselage?

Rather soon, the idea of flying such a big plane for our group alone began to look somewhat unreasonable, not just to me, but also to the authorities. It was clear that we should make this an opportunity for other astronomers, in fact, as many as the aircraft could safely carry. But how should they be chosen? Another French team stood out. At 32, Serge Koutchmy was already an accomplished eclipse hunter. He would eventually become one of the best specialists in the world for observations of the solar corona. Astronomer at the Paris Institute of Astrophysics (IAP), he was thus preparing for the eclipse of the century by planning two expeditions, one in Mauritania and one in Chad. These would prove extremely successful, allowing him to obtain first rate colour photographs of the corona (see Fig. 1.1). He was thus unavailable for the Concorde flight, but put forward a carefully considered corona observation programme based on techniques he knew well, which were thus less adventurous than our own infrared programme. He designed an instrument that would complement his ground-based projects and delegated one of his colleagues, Jean Bégot, to set it up in the plane.

In the United Kingdom, manifesting the close collaboration between the two countries, another prototype by the name of Concorde 002 had been built, in particular to develop the engines built by Rolls Royce. However, it would only make its first flight on 10 January 1973, 13 months after the maiden flight of Concorde 001. The British astronomer John Beckman, himself interested in solar infrared radiation, had recognised Concorde's potential for observing the eclipse and had approached the British Aircraft Corporation (BAC) which had built the 002 prototype. The response was negative. The company was concerned about upsetting its schedule, and in particular about creating a new porthole in a plane that had not yet even started its test flights. Although I had been quite unaware of these goings on, I already knew of John's scientific work and greatly appreciated his experimental abilities. We thus offered him a place aboard 001, and he accepted. His scientific programme would study the narrow zone immediately around the rim of the solar disk, a region known as the chromosphere. The name derives from the Greek word *chromos*, meaning colour. Indeed, the light emitted by the hydrogen present in this very thin gaseous layer is a beautiful red colour, very briefly visible during an eclipse at the instant when the lunar disk, just as it masks the Sun, allows a glimpse of this thin band of light. But what lasts on the ground only for a brief instant would last ten times longer with the help of Concorde, slowing as it does the vast sweep of the Moon's shadow. John, who hoped to study this poorly understood zone, would thus have much longer to measure its radiation. He, too, would benefit from the stratospheric flight to gain access to an infrared radiation that is practically impossible to observe from the Earth's surface. Hence, in 1966, prior to John's project, an expedition by a few young American colleagues who would subsequently become highly reputed astronomers had been to the region of Arequipa in Peru, at an altitude of 4500 m, to try to detect this radiation from the chromosphere, although with somewhat limited success.[4]

So that was already three holes that would have to be made in the fuselage. They would be made in the ceiling of the cabin since, during the eclipse, at the summer solstice and practically on the Tropic of Cancer, the Sun would be close to the zenith, that is, vertically overhead. We wanted them to be the maximum authorised size allowed by safety considerations. Indeed, the special transparent windows to be placed in these portholes would have to resist the pressure difference between the cabin and the outside, but also the high temperature—around 100 °C—reached by the outer surface of the fuselage when in supersonic flight, simply due to friction with the surrounding air. If the porthole were to explode in flight, although it would not be a major disaster for the plane and its passengers, it would nevertheless mean an emergency landing and a sudden end to all our observations. The Aérospatiale research unit generously authorised a further hole to be made (Figs. 4.2 and 4.3), so we were able to invite the American astronomer Donald Liebenberg aboard, another specialist in the study of the solar corona who, with his flight over Samoa in 1965, had already written a new chapter in the history of airborne eclipse observation. He belonged to the long line of pioneers that began with Mahieu in 1912, then Stevens in 1937, Michard in 1955, and several others after them.

In 1963, spelling the end of what was undoubtedly the most frightening period of the Cold War between the Soviet Union and the United States, an international treaty was signed which prohibited nuclear tests in the atmosphere. To monitor explosions carried out by the other side by flying at high altitude and outside the borders of the Soviet Union, the US Air Force had equipped three subsonic four-jet aircraft (NC-135) with suitable detection systems, and these were suddenly freed up as a consequence of the treaty. One of these was subsequently dedicated to the scientific study of solar eclipses.[5] The young Donald Liebenberg took part in the expedition which went to Pago Pago in Samoa in 1965 and obtained excellent in-flight images of the corona during almost three and a half minutes of totality. Just prior to this, the plane had come upon some annoying high altitude clouds and the pilot had had to use a rather hurried emergency procedure, normally only allowed in

---

[4] R. W. Noyes, J. M. Beckers, F. J. Low: Observational studies of the solar intensity profile in the far infrared and millimeter regions. In: Solar Physics **3**, 36–46 (1968).

[5] W. W. Dolci: Milestones in airborne astronomy: from the 1920s to the present. American Institute of Aeronautics and Astronautics, SAE Transactions **106**, no. 1 (1997).

**Fig. 4.2** In the spring of 1973, in the Aérospatiale hangar in Toulouse, four extra portholes were made along the axis of the Concorde 001 cabin (*arrows*). It would be through these openings that the four teams of astronomers would observe the solar corona on the day of the eclipse. The fifth experiment would use an already existing side opening. To check the adequacy of these portholes under the pressure difference generated by a stratospheric flight, the plane was literally filled up with compressed air in the hangar. © Aérospatiale/courtesy Pierre Léna

time of war, to climb higher without delay.[6] Donald had then taken part in all the NC-135 missions in Brazil, Texas, and Canada. It was thus natural enough that we should invite him to fly aboard our 001 so that his instrument could benefit from the fourth porthole, and that we should provide him with a lighter folding chair than an ordinary seat so as not to overload the plane at take-off.

[6] B. Mulkin: In flight: the story of Los Alamos eclipse expeditions. In: Los Alamos Science, winter/spring 1981, p. 39 ff.

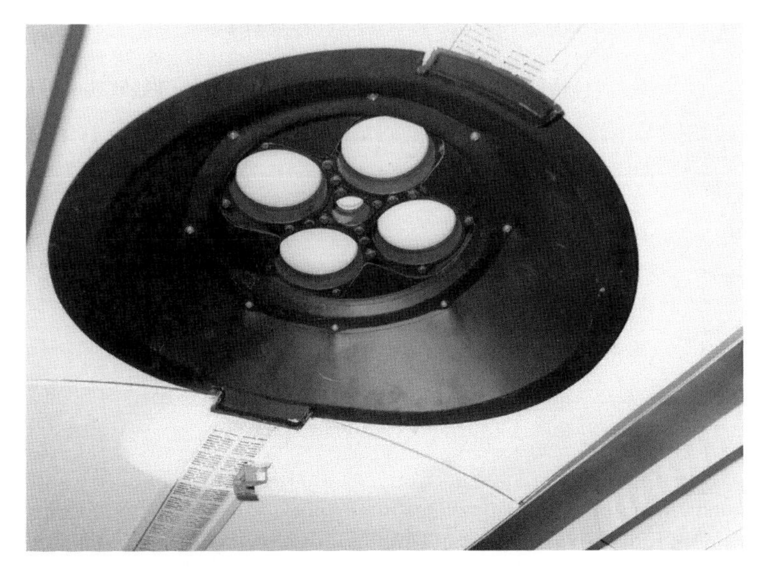

**Fig. 4.3** The porthole cover in the form of a four-leaved clover, seen from inside the cabin. Through this the instrument from the Paris Observatory would be able to view the Sun. It is still visible on the prototype Concorde 001 exposed at the aerospace museum in Le Bourget, France. © Aérospatiale/courtesy Pierre Léna

## Some Very Demanding Astronomers

In close relation with the astronomers and engineers who would accompany us, the Aérospatiale research unit was working on the problem of the portholes, the electricity supply for the instrumentation, the authorised mass of the aircraft, and even the number of seats possible in the remaining free space, which would determine how many people would be lucky enough to fly with us. We could already tell that seats would come at a price and I was careful not to make promises to anyone.

But the flight? Would it really come off? Let us hear once again what André Turcat had to say:

> I immediately entrusted this preparation to Henri Perrier, the chief navigation engineer, faithful since the first flight. For me there remained above all the preliminary question: the decision about whether it was wise to land at Fort Lamy at around 2 p.m. on 30 June, a bad time for meteorological reasons. What would be the fuel reserve upon arrival if I had to wait for a storm to pass? And within the framework of all our work aimed at test flights, we would be able to prepare and include our eclipse, each of us concerned with the need to be thorough and with the desire to be conclusive and not to miss out on this extraordinary opportunity. But there would be no point in getting enthusiastic about something that turned out to be impossible, would there?
>
> And to begin with, what should be our starting point? Dakar? Perrier had quickly ruled that out. The runway was not very long, and well cleared of obstacles, but above all, it would be hot at the date and time of take-off, and this would reduce the maximal allowed mass at take-off, that is, the amount of fuel. Furthermore, the weather conditions could be poor in the harmattan or dust haze.

I suggested Sal, an island in Cape Verde, where we had already made a stopover on the way to South America. It was a good runway with very regular weather conditions, always north of the intertropical convergence zone, but it would be difficult to lodge any large number of people, for there was no water on the island, simple bed and board but no hotel, and no regular air link.

Perrier came up with Las Palmas in the Canaries. There were none of the disadvantages of Sal. The weather was less reliable but never bad, the temperature was a little higher but never above 26 °C. However, Las Palmas was a little further from the rendezvous and would require take-off in the wrong direction, into the trade winds. Perrier showed me that it was rather an advantage, avoiding a costly U-turn during the climb. We discussed all this. A more precise calculation of the details would decide, so we sent all the facts to the computing unit. The result came back the next day: 2 or 3 tonnes more for Las Palmas. Perrier was right, as always. And provided that there were no modifications to the details of the flight path by the astronomers, I would have 10 tonnes of kerosene upon arrival in Fort Lamy, forty minutes' wait, and the right to a missed approach. It was reasonable, but a storm can last an hour, and imagine an approach in very poor weather conditions if one is running out of fuel. Consulting a booklet on the local weather, I was pleased to learn that, in this period at the beginning of the rainy season, storms rarely occur before 4 or 5 p.m. local time. Further, a phone call to colleagues at the company UTA, familiar with the area, confirmed this information and told me more, invaluable this time: it was exceptional for Fort Lamy and Kano in Nigeria to be blocked at the same time. Kano had a good runway where I had once landed a military Dakota aircraft. That would leave me the possibility of a rerouting decision up until the moment of descent. From A. Turcat, Concorde : essais et batailles, p. 244–248. See bibliography.

Astronomy also imposed constraints on the Concorde flight path, although of another kind. The geometrical features of the eclipse raised no difficulties. The exact position of the centre of the Moon's shadow on the surface of the Earth, or at around the altitude of 17,000 m at which we would fly, the exact size of the shadow which strictly specified the region in which the Sun would be totally hidden, and the speed at which it would move relative to the ground were all perfectly computed and freely available in tables published for each future eclipse. In my notebook of the time, I find my first estimate of the maximum duration of totality, based on a possible speed of the aircraft and in the absence of wind: 83 min and 58 s! But for this, the pilot must first have mapped out the ideal flight path and inserted the details into the onboard computer, informed by the two onboard inertial guidance systems. Their gyroscopes hold a fixed position relative to the stars and can be used to determine the geographical location on the map at any time, hence also the distance covered since the point of departure of the flight. Any inaccuracy in these systems would lead to an error in the rendezvous, any breakdown to a missed rendezvous, whence the aircraft had two, working independently.

In the United States, Donald Liebenberg had all the tools needed to carry out the necessary calculations for an observation from a rapidly moving aircraft. He suggested a flight path made up of three arcs of a great circle. The advantage here was that each of these three 'straight lines' plotted on the terrestrial sphere would allow the plane to fly without bends, simplifying the pilot's task and also the accurate pointing of our instruments. In the end, from Donald's calculation, we chose a path made up of a single arc of a great circle, more favourable for carrying out observations because there was no bend in it at all (Fig. 4.4). It would touch the

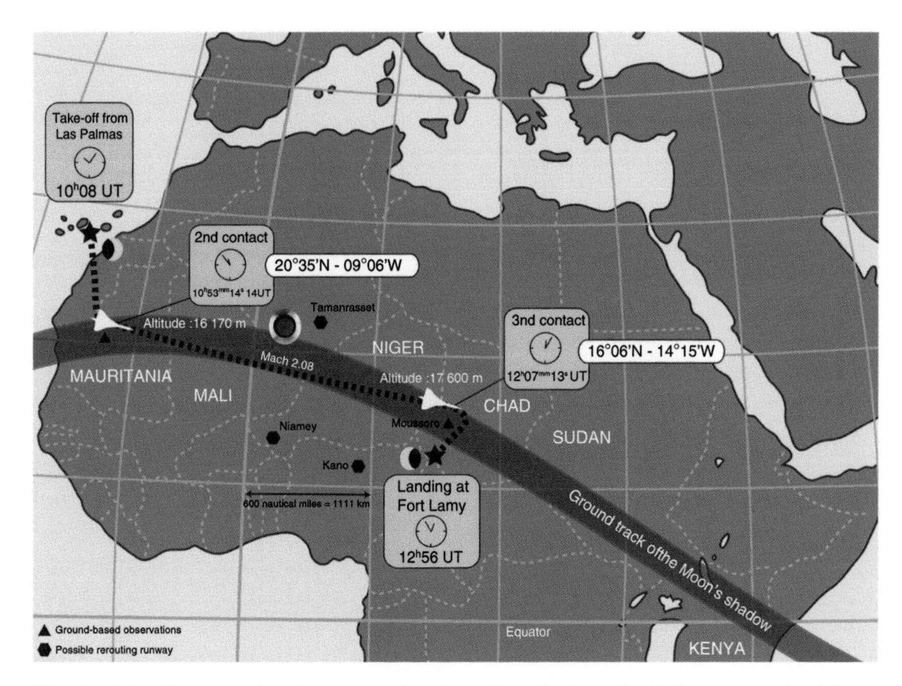


**Fig. 4.4** The flight plan for 30 June, designed to chase the Moon's shadow across the African continent. Total darkening of the solar disk occurs between the second and third contacts, those instants when the lunar limb and the solar limb touch one another. First and fourth contacts refer to the beginning and the end of the eclipse phenomenon. © G. Zimmerman

edge of the shadow at the instant known as second contact, when the solar disk suddenly becomes invisible, cut directly across it to reach its southern edge, then as the shadow overtook us, leave by its northern edge, after which we would come to land. If there were no opposing wind, this path would give us 80 min of darkness. Demanding as ever, we also required this rendezvous with the shadow during second contact to occur to within 15 s of a fixed time, at a fixed geographical point over Mauritania. We argued for maximal altitude. That was no problem because Concorde flies higher when the atmosphere is colder. But close to the equator, the stratosphere is colder than at medium latitudes and we could count on a favourable average altitude of 17,000 m. We also wished to slightly exceed Mach 2 in order to follow the shadow for as long as possible. The authorised speed limit for the 001 prototype was Mach 2.05. However, if the atmosphere is colder, the speed of sound (which depends on the air temperature) is lower there, and Mach 2.05 would give us a lower speed relative to the ground, hence also relative to the shadow, thus reducing the possible duration of totality.

All this was communicated to André Turcat so that he could prepare the flight schedule down to the slightest detail. On 2 February, the decision was taken by the authorities: we would fly!

## Hands on

On my desk, I had long since drawn up the list of ingredients that must be incorporated into our observing instrument, and the list of constraints that had to be satisfied: limited space in the narrow aircraft cabin, weight, and electricity consumption. For safety reasons, we had to be sure that nothing would move in the case of an emergency landing or severe turbulence, and also that there was no risk of a short-circuit that might start a fire. A list of somewhat disparate items[7] was soon ready: a porthole made from an exotic and costly material called Irtran, which would allow infrared light to pass through; a large gyro-controlled movable mirror mounted on rails, which could accurately reflect the light whatever the residual motion of the plane or the position of the Sun relative to the plane (and this would change significantly as we were flying to the east); to make an image of the Sun and its corona, a telescope that would reduce to a large and light aluminium dish, not too costly; and finally, the key feature, three ultrasensitive 'thermometers' to detect the infrared light emitted by the corona. Back in Arizona, Don carefully prepared two of these 'thermometers'. With a certain daring, he proposed a new kind of instrument that he had just developed. The eclipse flight would be its first astronomical application, but later this device, made from two rather exotic chemical elements, indium and antimony, would be widely used in many observatories around the world. Regarding our third 'thermometer', I owe this to the young French scientist Noël Coron and his skilful technical team. A brilliant experimental physicist but sometimes tough to deal with, he imposed drastic conditions on the loan of his equipment, but as an amateur pilot himself, he could hardly have done otherwise and accepted gracefully. These three 'thermometers' could only work correctly at unbelievably low temperatures. To keep them cool in flight, we had no choice but to use liquid nitrogen and helium. These precious cooling fluids would have to be supplied in Toulouse and taken to Africa in large Thermos flasks. If by bad luck they were to evaporate too quickly, our measurements would become impossible, and so also would John's, for his instrument would also depend on it. Finally, we would need to add a touch of electronic circuitry to our list to get the whole thing working; a pump to lower the temperature of our 'thermometers'; two large recording devices, one with pen and ink and the other with magnetic tapes, which would store our results and which were generously loaned to us.

All we had to do was put the whole thing together. My two engineering accomplices at the CNRS, Charles Darpentigny and Alain Soufflot, worked marvels, always calmly and in good humour, trying for my own impatience, for the deadlines were coming up fast. The first had a long experience in aeronautics, having started his career in Nord Aviation and then designed our telescope on the Caravelle (see Fig. 3.4). The second had already been entrusted with the task of building a French instrument for solar exploration which would be sent into space

[7] The French would call it a list *à la Prévert*, with reference to Jacques Prévert's famous poem *Inventaire*, which lists all sorts of items with no apparent relation between them.

by NASA in 1975, aboard the Orbiting Solar Observatory (OSO-8). Drawing, cutting and milling, mounting and aligning, plugging in wadges of multicoloured electrical wires, everything fell into place with a healthy mix of do-it-yourself and professionalism, while keeping an eye on a budget which had little elasticity and on the weight which could not exceed 400 kg. By the month of April, we were able to send our strange machine to Toulouse and fit it aboard the 001 to see if it would work correctly (Fig. 4.5). In the huge hangar that housed the aircraft, there was a persistent smell of oil which, like one of Proust's celebrated madeleines, took me straight back to those nights in California 5 years earlier, when I had stood next to NASA's four-jet Galileo I aircraft, my first encounter with the world of aviation.

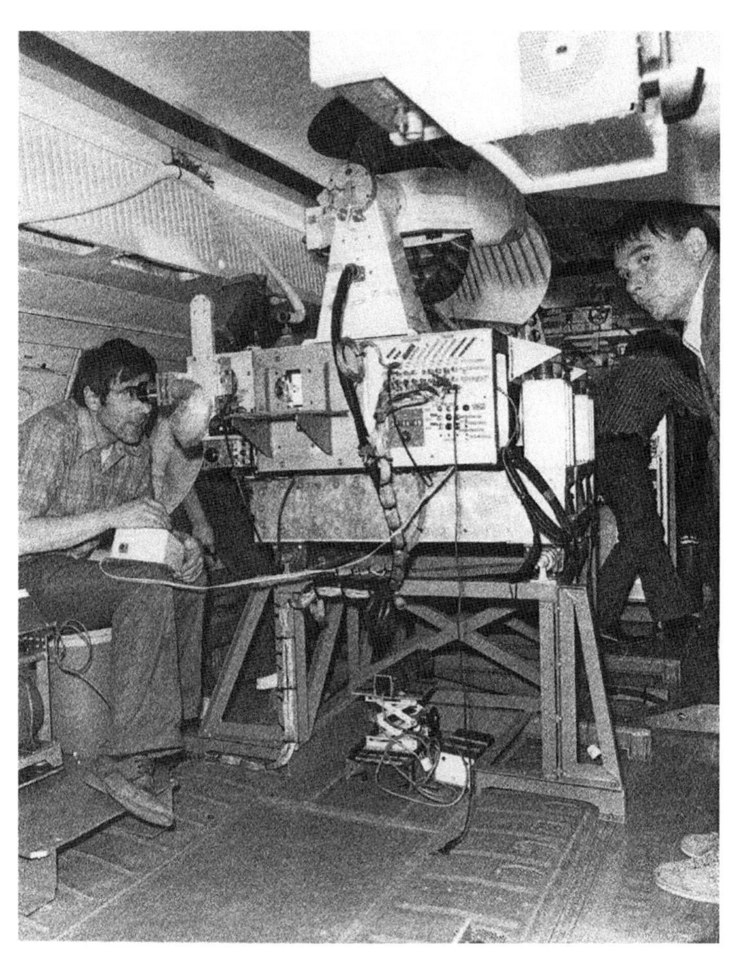

**Fig. 4.5** The instrument built at the Paris Observatory and the prototype department of the CNRS is placed vertically below the porthole. The author (*left*) is checking the optical alignments. The engineer Charles Darpentigny (*right*) designed the instrument, taking great care to respect aeronautic safety standards. © Pierre Léna

One afternoon in Toulouse, we were alone in the plane fixing up the circuitry when a strong smell of burning suddenly came to our notice, causing general alarm. We were not proud of ourselves, until the plane's technical team noticed that the short circuit had occurred in a quite different location and thereby cleared us of responsibility for any false manipulation. So work went on, and in the meantime our British and American colleagues were each setting up their equipment. In fact, a fifth partner, Paul Wraight, had come from Scotland to join our conspiracy. This physicist did not require any further hole to be made in the cabin, for he was not planning to observe the eclipsed Sun, but rather the effect produced by sudden darkness on oxygen atoms in our terrestrial atmosphere. To do this, he needed only to aim slightly above the horizontal, through a normal porthole whose pane had simply been replaced by a transparent and more carefully polished crystal window.

## Atmosphere and More Atmosphere

Week by week, in the hangar in Toulouse, the team's excitement was mounting, and all the while the deadline was moving closer. Fortunately, the plane was not required to fly too often. But why was the 001 prototype, originally programmed to end its test flight career at the end of December 1972, allowed to go on for a further full semester? Was it for the exclusive but costly benefit of a handful of lucky astronomers? Not exactly.

Apart from the visible light perceived by the retina in our eyes and the infrared radiation picked up by our skin cells, the Sun also emits ultraviolet (UV) light which, in small doses, tans the skin of bathers and mountain dwellers, but at the risk of sometimes dangerous melanomas. But in larger amounts, this UV radiation could wipe out all forms of life on Earth. Fortunately though, above an altitude of 15 km, the stratosphere contains a gas called ozone, derived from the usual dioxygen molecules $O_2$. Molecules of ozone each contain three oxygen atoms rather than just two. This blocks out most of the solar UV and thus plays the role of an invaluable protective filter. But how fragile is it? Given its crucial importance, the question had long been raised by chemists, but it was not an easy one to answer. Indeed, it was often enough for the molecules to be present in very small amounts in order to upset the complex chemical equilibrium which had maintained this atmospheric filter in place over the centuries. Some of these molecules had been identified in quite an unexpected way. In 1942, in Belgium under German occupation, professor Marcel Nicolet had been ordered to interrupt his meteorological studies. He subsequently took up an interest in the chemical equilibrium of the stratosphere and discovered the role played by certain molecules, in particular, oxides of nitrogen, in this equilibrium.

Supersonic aircraft fly in the stratosphere, much higher than subsonic airliners, because it is at this altitude, where the air density is some ten times lower than at sea level, that aerodynamic performance at Mach 2 and engine performance are at their best. And it turns out that the exhaust gases from the engines contain nitrogen oxides. So at the beginning of the 1970s, environmental activists were concerned about the possible harmful effects for the ozone layer that might be caused by a large commercial fleet of supersonic aircraft. Such a question could not be left unanswered and research began, significantly improving our understanding of the subtle stratospheric chemistry. But measurements had to be made directly in the stratosphere, and naturally, Concorde 001 offered an ideal platform for setting up the necessary heavy equipment. The latter, viewing through a lateral porthole of the aircraft, picked up the infrared light emitted by atmospheric nitrogen oxide molecules excited by solar light. Invaluable measurements were thus obtained regarding the quantities and reactivities of these molecules. Using these measurements and the amounts of gas ejected by the engines, chemists could then determine whether there was any significant risk. In this way, by the end of 1972, all the science goals, those of atmospheric chemistry and those of astrophysics, had been successfully brought together, whence the political decision-makers elected to maintain the prototype in flight through the first part of 1973. This was an unexpected and lucky convergence, which I gradually discovered during the friendly common meetings in which we set up the flight schedules.

One colleague, the austere but generous physicist André Girard, had long been pioneering the study of infrared light. The French government, extremely attentive to the possible harmful consequences of these notorious oxides, had put him in charge of part of the study because he had designed a remarkably powerful and novel analysis device. Viewing the Sun low on the horizon through the Concorde porthole, this device analyses the Sun's light, part of which has been absorbed by the thick layer of the Earth's atmosphere it has had to cross to get there. It thus reveals the presence of nitrogen oxides. Science flights by 001 equipped with this instrument, in which André Girard took part to study the stratosphere, were thus carried out in the spring of 1973, alternating with tests of our own instruments in the hangar in Toulouse. Another colleague, André Marten from the former National telecommunications research centre (*Centre national d'études des télécommunications* CNET), who would later join our laboratory in Meudon, had been specialising for some time in the study of planetary atmospheres, and particularly those of Jupiter and Saturn. So why not apply his research techniques to this problem of the terrestrial nitrogen oxides? He thus came to test his instruments aboard the Caravelle 116 at the flight test centre, before setting them up in Concorde 001 in this same spring of 1973.

On 17 May, we were delighted to hear that we would be able to test the behaviour of our instruments during a flight devoted to the study of the stratosphere that would take us from Toulouse right out over the Atlantic. This would be our first flight aboard 001. From my viewpoint during take-off, I could see the long pointed nose of the aircraft, starting in the lowered position and then lifted up in flight, while a visor system moved into place to obstruct part of the view from the cockpit when the plane accelerated. The large orange circle on the radar screen showed us the Bay of Biscay, which we soon left behind to break the sound barrier over the ocean, not wanting the supersonic 'boom' to disturb the duck farms in the south west of France. Moving faster than sound in this westerly direction "brought the Sun back up above the horizon behind which it had just fallen", as our chief pilot André Turcat so nicely put it.[8] For the first time in my life, I was going faster than sound, and yet there was nothing to show that this was the case. It was a strange sensation, one that thousands of Concorde passengers would later experience on their way to New York.

Even though there was no eclipse, the navigation had to be accurate. The plane's direction had to be held at 90° to the direction of the Sun, which was close to setting. The thickness of the atmosphere crossed by the Sun's light absorbs certain colours, in particular blue, as we see every evening when the Sun and sky turn an orangy red, the colour complementary to blue. Stratospheric nitrogen oxide molecules also absorb this solar light at certain precisely defined wavelengths, and it was the intensity of these wavelengths that the onboard measuring devices were designed to measure (Fig. 4.6). Once the measurements were made, we returned to Toulouse, while the great veil of night climbed to the east, the Earth's gigantic shadow on the pure atmosphere, a shadow I had so often admired from the observatories in Arizona and Chile, before my dialogue with the stars. But here, in the perfectly clear terrestrial shadow, in a sky of unequalled purity, the horizon was curved, for at that altitude, one can witness the Earth's roundness directly.

The pilots, who showed a friendly respect for the rest of us, were very pleased with the first flight made in our company, whose navigation accuracy augured well for things to come. With his usual faultless vigilance, André Turcat was surprised by the way our delicate equipment stood up to the vibrations during the flight, and he congratulated us for that. Charles, who designed this feature, smiled modestly. We had just had our maiden supersonic flight, before a second rehearsal in May. We would have to dismantle everything to free up space in the cabin, then put it all back in a few days just a month later.

The results of these studies of the stratosphere were eagerly awaited, but it was not until 1976 that the thick report by the stratospheric flight committee set up by the French government was finally published. And it turned out that stratospheric flights would have an effect on the protective ozone layer. But the measurements and calculations showed that the impact of a hundred Concordes in regular service would affect this layer by barely 1 %, while the natural annual variability was already 20 %, because ozone chemistry is also sensitive to other factors, such as the alternation of day and night or the 11 year cycle in the Sun's magnetic activity. So there was nothing to worry about! But science sometimes makes unexpected detours. Just as German occupation of Belgium had led Nicolet to investigate stratospheric chemistry, fears about the impact of supersonic flights made some look more closely at the behaviour of this ozone layer. And in this same year of 1976, the publication of a report by the National Academy of Science in the United States caused new alarm. From earlier measurements made over a period of more than a decade, the report showed a constant and worrying decrease in the ozone layer that would be confirmed 3 years later by the formation of an 'ozone hole'

[8] A. Turcat: *Concorde: essais et batailles*, p. 238. See the bibliography.

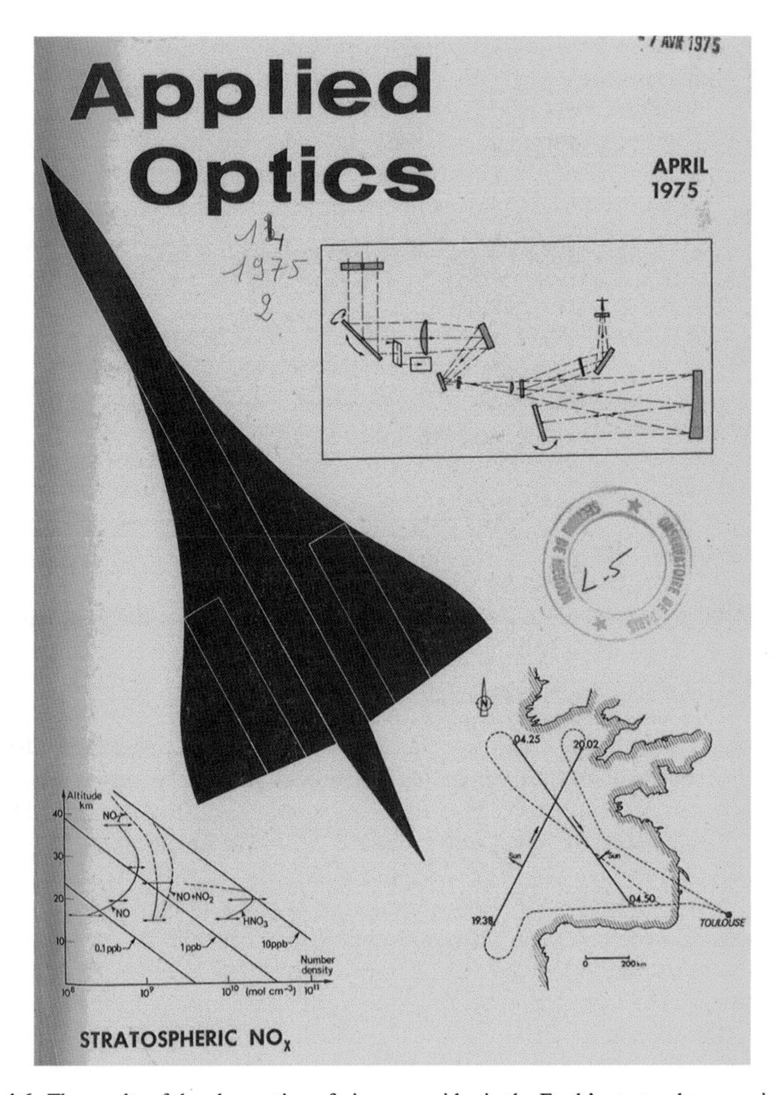


**Fig. 4.6** The results of the observation of nitrogen oxides in the Earth's stratosphere, carried out by André Girard on Concorde 001, are featured in this issue of the scientific journal *Applied Optics*, which appeared in 1975. The consequences to be drawn regarding possible impacts on the ozone layer of a fleet of supersonic aircraft flying in the stratosphere would be the subject of an official report in 1976. © Optical Society of America (OSA)/Journal Applied Optics, front cover from April 1975

above the Antarctic continent. The reason for this decrease had been suggested back in 1974 by the American chemist Frank Sherwood Rowland, who had shown that chlorine-and fluorine-containing molecules used in vehicle air conditioners and private and industrial refrigeration systems, then released in large amounts into the atmosphere as a gas, were gradually destroying the ozone molecules.[9] Two years after the publication of the report, the United States and several other countries prohibited the production of these so-called chlorofluorocarbons (CFCs). Then in 1987, in an exemplary move, unfortunately rare at this level, the international community of 43 represented countries signed the Montreal Protocol which banned the use of such molecules. Sherwood Rowland (USA), Paul Crutzen (Netherlands), and Mario Molina (Mexico) shared the 1995 Nobel Prize in Chemistry for their work on atmospheric ozone chemistry.

It would be a good thing if, faced with the increasing concentration of carbon dioxide in the Earth's atmosphere, the major risks associated with climate change would lead today to such radical and courageous measures on the part of the international community, for it challenges many private interests.

## Ready for Service

Four months had gone by since the green light of February 1973. Following confirmation of the flight plan, André Turcat obtained all the overflight permits required, particularly for the day of the eclipse, where we would have to fly over the Sahara, then Spanish (but administered by Morocco in 2014), Mauritania, Mali, Nigeria, and Chad, before landing in the latter. Only Mauritania refused at first. Indeed, this country was to host astronomy expeditions from the world over and had decided that, on 30 June, this part of the African sky would be closed to all air traffic. What a disaster it would have been if a plane had unexpectedly come between the eclipsed Sun and a telescope, or worse still, if a beautiful condensation trail (contrail) were to float by during a measurement made during totality! That is, unless some scientific sounding rocket should happen to collide with Concorde! But having convinced the authorities that Concorde's route had been determined with all the necessary precautions with regard to ground-based observation sites, thus reassuring our colleagues who would be using those sites, Mauritania finally gave us the permit (Fig. 4.7).

Another of my astronomer colleagues, Pierre Charvin, held a position of responsibility in the CNRS. From the very beginning of the project, he had been in charge of coordinating all aspects of organisation and flight paths, thus relieving me with great skill from a task that I could not have done properly at the same time as preparing my own experiment aboard the plane. Aérospatiale had also allocated an enthusiastic and cautious engineer, René Joatton, from the Burgundy region like myself, to follow the scientific projects aboard Concorde and serve as the counterpart to Pierre for the eclipse mission. Diplomacy, insurance, and the press were thus admirably handled from start to finish. As head of mission, Pierre flew with us on

[9] http://en.wikipedia.org/wiki/Ozone_depletion

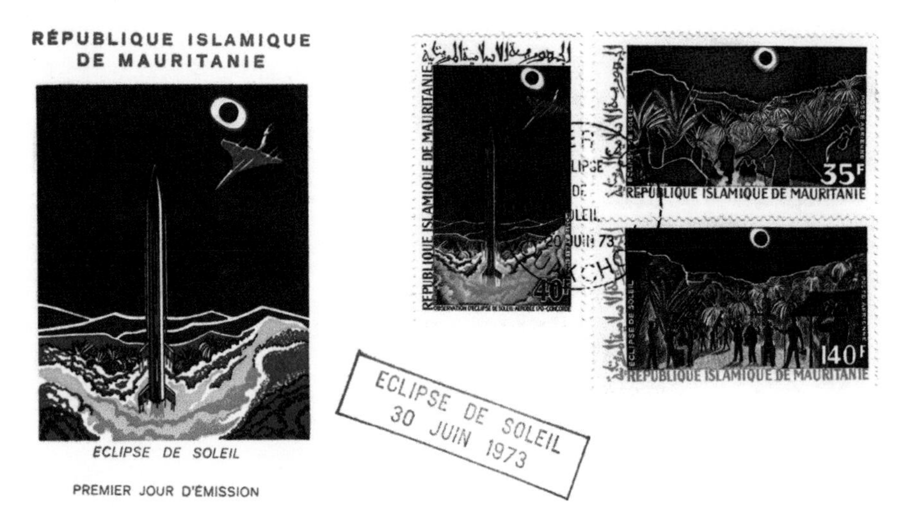


**Fig. 4.7** This first day cover, of a kind so dear to philatelists, shows different scientific observations, on the ground, aboard Concorde 001, and using a sounding rocket, all carried out in Mauritania during the total eclipse. Document kindly provided by Henri Aubry

the big day, as did his assistant engineer Michel Ravaut. Years later, he would become the president of the Paris Observatory and we would sometimes reminisce about our shared supersonic experience. On 23 June, before leaving Toulouse for Africa, we held the first real rehearsal for all the astronomers and their instruments. A path was calculated that would take us far down to the south. Everything worked, except for our paper registering device, in which the pens became blocked, for in this distant period, such recording devices still used ink pens. Things could have been worse. We were ready for the real thing and took off for the Canary Islands on 27 June. A Caravelle came with us for technical support. It carried the large 'Thermos flasks' containing our crucial and precious cryogenic liquids: dozens of litres of liquid nitrogen at minus 196 °C and liquid helium at −269 °C. It also carried a replacement engine for the 001, and all the technical teams for whom there was not room enough aboard the supersonic aircraft, and who would not therefore experience the breaking of the sound barrier.

Concorde landed at Las Palmas and excited much curiosity (Fig. 4.8). The Spanish Guardia Civil watched over it day and night.

"CONCORDE": A LA SOMBRA DE LA PIEL DEL SOL

A BORDO DEL AVION SUPERSONICO, SABIOS DE VARIOS PAISES ESTUDIARAN EL ECLIPSE DE SOL DEL SABADO

Un vuelo histórico de 80 minutos: desde Gran Canaria a Fort Lami a 2.100 Kms.-h. y a 18.000 m. de altitud

—(INFORMACION EN PAGINA 13)—

En este mapa se aprecia el recorrido del «Concorde» de Las Palmas y se introduzca en la sombra del eclip Mauritania para abandonarla a su paso sobre For República del Tchad.

EL ECO DE CANARIAS

EL PERIODICO DE LAS ISLA



**Fig. 4.8** Two days before the eclipse, the Las Palmas press marvelled at the whole affair with the somewhat surprising headline: "In the shadow of the Sun's skin". Perhaps they were referring to the chromosphere? © El Eco de Canarias, documents obtained from the Jable Digital Press Archive at the University of Las Palmas de Gran Canaria (ULPGC)

# Chapter 5
# No Need for Alarm, It's Only an Eclipse

Such was the calm declaration of Professor Calculus, although about to be burnt at the stake, as the Andean landscape suddenly went dark in *Prisoners of the Sun*, one of the adventures of Tintin (Fig. 5.1). On the morning of 30 June, as the Sun rose, already dented by the Moon, we were less composed than the celebrated savant with his pendulum, for any mishaps, no matter how small, might threaten the success of the whole venture.

## General Rehearsal

Two days before the event, on 28 June, we did a final rehearsal flight, taking off from Las Palmas and this time following precisely the flight path scheduled for 2 days hence, and even respecting the rendezvous times. Aboard the plane, each instrument was managed by a single astronomer. Stationed by his porthole, he had therefore to deal with every issue that came up. Given the complexity of our own instrument, I was allowed three people to manage it: Alain Soufflot, Don Hall, and myself, proudly wearing our bright red flight suits. I did have to sort out a slight hiccup: for some reason, the technician who fitted our recording device, a stubborn chap from Britanny who came in the Caravelle, had imagined that he was going to fly in the Concorde. I had to set him straight and his reaction was somewhat churlish. I had to be extremely diplomatic in order to get the final adjustments of the device. These were crucial to us and he was the only one who knew them! Before taking off, standing below the great white bird under the tropical Sun, Charles Darpentigny and Jacqueline Mondellini helped us to pour the cryogenic fluid into the small and delicate cryostats containing the 'thermometers', hypersensitive to infrared radiation. John Beckman did likewise. We set them up on the specially built construction aboard the aircraft, started the pumps, checked the



P. Léna, *Racing the Moon's Shadow with Concorde 001*, Astronomers' Universe,
DOI 10.1007/978-3-319-21729-1_5

**Fig. 5.1** While the dog Milou barks, Professor Calculus quietly comments *No worry ! This is simply an eclipse.* Was Tintin's adventure in *Prisoners of the Sun* inspired by a genuine eclipse above the land of the Incas? Serge Koutchmy and Robert Mochkovitch investigated the question and concluded that the eclipse of 25 January 1944, crossing South America from west to east, was a possible candidate. However, at the given time, it could not have affected anywhere inhabited by the Incas. (P. Guillermier, S. Koutchmy: *Eclipses totales: histoires, découvertes, explorations.* See the bibliography). © Hergé/Moulinsart, 2014

alignments, and tightened the clamps that would hold all this mechanics in place despite the vibrations that would occur during take-off. The acceleration was surprisingly sudden, strong enough to hold us to our seats while the four jet engines roared and the afterburn tore us away from the ground.

Today the crew of six was precisely the one that would fly 2 days later. The pilots, André Turcat and Jean Dabos, were at the controls, while Michel Rétif, the test flight mechanic, was seated immediately behind them, keeping a close eye on everything, and especially the engines. Further behind was the test facility, a long row of cabinets packed with electronics and specific to the 001 prototype. There was a console on the cabin axis. Opposite these rows of dials and indicator lights were seated Henri Perrier, the test flight navigating engineer, Jean Conche who monitored the jets and their oil consumption, and Hubert Guyonnet controlling the navigation, all of them dressed in red. Through our headphones, we heard the exchanges between them on the onboard radio system. We were accompanied by Jean-Pierre Aubertin, photographer and filmmaker. He was the only one whose presence was accepted by André Turcat during our flights, despite the many requests that were presumably made. Smiling and as discreet as a mouse, he produced the magnificent images that would eventually be made into the only film of the event, entitled *Eclipse 73*.

The stratosphere was still poorly understood, especially above Africa, where meteorological sounding balloons were rare and where observation satellites were still not available. Over the path that we would follow in pursuit of the Moon's shadow, the rehearsal flight of 28 June provided accurate measurements of the air temperature as a function of altitude. This was invaluable information for refining

**Fig. 5.2** After the rehearsal flight of 23 June, standing by Concorde 001 on the tarmac at the airport in Blagnac near Toulouse, the pilots André Turcat and Jean Dabos (wearing flight suits), together with the science team. John Beckman proudly carries the delicate 'Thermos flask' which had contained liquid helium, cooling his infrared-sensitive 'thermometer' to a temperature of around minus 269 °C. © F. Claudel, DR

the schedule that the crew would specify to obtain a precise rendezvous with the shadow. In André Turcat's words[1]:

> Take-off into the trade winds, retraction procedures for the landing gear, extinction of engine afterburns, half-turn toward the south, subsonic then supersonic climb, stabilisation at a speed just above Mach 2, swing to the east to connect with the great circle (orthodromy) at the stipulated point on the map and within the expected quarter of a minute, all in half an hour cut into sections numbered in seconds. For the flight two days later, we had already decided to take off twenty seconds earlier in order to make up for atmospheric irregularities, and during this rehearsal on 28 June, I devised a procedure that would allow us to accurately drop a few seconds by applying the small airbrakes that can be used at Mach 2. These had been fitted on the prototype at my request right from the start. They had to be deployed for ten seconds in order to lose one second, and this without interfering in any way with the engines, so as not to complicate the maneuver. We got back without mishap, nothing to modify aboard: we just had to await the big day.

When we touched down after 2 h and 36 min in the air, we were relieved for our instrument had worked well. We would just have to abandon one of the three infrared measurement channels whose 'thermometer' had been contaminated when the Thermos flask (cryostat) containing it had been opened. Nothing dramatic (Fig. 5.2). Our colleagues also seemed reassured, but we were concentrating so much on what would soon happen that we barely exchanged impressions during dinner at the hotel that evening.

[1] Op. Cit. note 29.

## Time to Go

The weather was stunning in Las Palmas on that morning of 30 June. The mild trade winds were blowing, causing the palm trees to wave, when suddenly a verse by Paul Verlaine came into my mind[2]:

*Le ciel est, par-dessus le toit,*
*Si bleu, si calme!*
*Un arbre, par-dessus le toit,*
*Berce sa palme.*

This was the sky in which we were about to fly, the sky it had become my job to study, raising as it does so many questions about what lies beyond. The rendezvous was somewhere above the town of Atar in Mauritania. The navigator, Hubert Guyonnet, had specified it as 20°49.4′N, 11°19.8′W, 10 h 46 min 00 s UT, in the inertial guidance systems that would be used to steer the plane. At each instant, their gyroscopes would be the only way of knowing and informing the navigator where the aircraft was located above the Earth's surface. Our filmmaker came aboard. The film sequences he would produce were so unique and beautiful that this narrative alone could not replace them. I can only advise my readers to view the film *Eclipse 73* as soon as possible, and live our flight sequence for themselves.[3]

Up in the sky, the eclipse had already begun, and a dark indentation was encroaching slowly on the solar disk. For an immobilised ground-based observer, the Moon moves from one side to the other across the Sun's surface in 4 h and 30 min. Everyone was taking place in the plane and carrying out their final verifications. With all the necessary precautions, we had begun operations half an hour ahead of schedule. One of the inertial guidance systems was not working and had to be changed. There was too much kerosene on board and some of it had to be burned off by running the engines to lighten the aircraft. Just a minute before the take-off signal, a small aircraft was given the go-ahead to fly! It was 10 h08 (on the Universal Time scale, used throughout the following): the plane took off exactly on time, after an about-turn and the time to move down the runway, respecting the flight schedule established 2 days previously to within a second. It augured well. Our rendezvous awaited us in 45 min and 14 s. The shadow hastened along beside us. Coming from the west across the Atlantic at a speed of more than 2200 km an hour, it would soon reach the African coast. I fixed an air navigation chart to the side

[2] The sky above the rooftop
Is so blue and calm!
A tree above the rooftop
Waves its palm.

[3] The film is called *Eclipse 73*. See the bibliography. http://www.youtube.com/watch?v=zHLyypLk-w and http://www.cerimes.fr/artciles/article_514/moment_771/plasma. See also the beautiful animated display of the 30th June flight and shadow trajectories at http://xjubier.free.fr/en/site_pages/solar_eclipses/TSE_19730630_Concorde001.html

**Fig. 5.3** The atmosphere aboard Concorde 001 during the eclipse flight, just before totality. In the foreground, Henri Perrier, the test flight engineer, is inspecting the map of Africa. Behind him is Michel Rétif, the mechanic, monitoring and checking all the flight systems. Further back, in the cockpit, darkened by the lifted visor on the aircraft nose, also lifted, are the pilots André Turcat (*left*) and Jean Dabos (*right*). © P. Léna

wall of the cabin. It showed the rendezvous, the time we should exit the shadow, and the predicted time of landing at Fort Lamy. I glanced at this occasionally to see how we were getting on.

In the words of our pilot once again[4] (Fig. 5.3):

> The atmosphere was less reliable than the previous day and made us lose eight seconds over the climb. We just had to apply the planned maneuver to lose the remaining twelve seconds by which we were ahead of schedule. Three minutes before the rendezvous, we were flying four seconds behind the ideal time, well within the already ambitious tolerance level, fixed at fifteen seconds by the astronomers. But a challenge is a challenge, and since it was not a commercial flight, I decided to increase the Mach number by a few points beyond the so-called Mach maximum operating (MMO), taking us close to the risk of 'pumping' in the engine air inlets. It sufficed to monitor the speed without a moment's inattention, but we

[4] Op. Cit. note 29.

> had already been doing that during the test flights. So in the end we came within one second and one nautical mile of our celestial rendezvous, not far from Atar, flying due east at Mach 2.04, with all the perfect stability of Concorde's characteristic cruise flight.

One or two minutes earlier, in the original viewfinder we had built ourselves, comprising a bundle of optical fibres and an eyepiece, I was watching carefully, now that the glare had become bearable to my eye, the image of the almost eclipsed Sun, located vertically above the plane, whose rays passed through the large porthole in the shape of a four-leaved clover, to be swallowed up in our instrument. The image-stabilising mirror, closely watched by Alain, worked extremely well as long as the plane was stable, with neither roll nor pitch. Suddenly, a red border appeared around the Moon's disk. This was radiation from hydrogen in the chromosphere. It was magnificent. Slowly, 15 times more slowly than for those observers now standing 16,170 m below, the lunar limb swept across the solar chromosphere and gave John Beckman the exact cross-section he had been hoping for. Not far from me, toward the back of the plane, Jean Bégot was guiding his telescope by hand and began to take a long sequence of photos. The instrument was hanging below the porthole in the cabin ceiling, and seeing him there clasping the long aluminium tube with both hands, he looked just like the lookout in a submarine, adjusting a periscope. I was dying to have a glimpse through the porthole, but there was no question of that, time was too precious. Concerning the photos of the spectacular events outside and the crew, totally focused on the task at hand, Jean-Pierre Aubertin was in charge of that.

The chromosphere disappeared. We reached second contact at exactly 10 h 53 min 14 s Universal Time. André Turcat had met the challenge quite magnificently, and Pierre made the announcement over the radio channel. In the plane and better than any ground-based observer, only two astronomers, Jean and Donald, could really enjoy the splendour of the corona through their instrument, when suddenly it came into view at the zenith, surrounded by stars, in a sky that was almost black at this high altitude. The crew and the other astronomers were unable to admire this stunning spectacle. For now began the infrared observation of this solar halo that had been seen by Plutarch and Thomas Edison. Fortunately, no other flying object would be able to disturb us. A few moments before the second contact, I had already checked that the sighting was adjusted to a short distance from the solar limb, in fact, 1.5° of arc, with a carefully calibrated ruler. I thus switched on the motor that would slowly scan across to produce an image of the ring of dust, whose presence or absence our two 'thermometers' was about to confirm. Over the radio, the recording of which would be crucial when it came to analysing the measurements, I spelt out for Don: "Pip alignment four [solar] radii. Pip five radii. Pip six radii. Pip seven radii. Restart: repointing on three radii. Pip . . . Pip for seven. Pip for eight. Excellent." Then a somewhat incongruous "Sh. . .t!' After the second contact, there was no longer any parasitic light from the solar disk and our 'thermometers' were lit solely by coronal light, and by the intense infrared radiation from our porthole heated up to more than 100 °C by the supersonic air friction. Of course, we had designed a device to subtract this rather annoying signal. We scanned back and forth across the coronal region close to the Sun in order to accumulate as many measurements as possible. A glance at the paper register (whose ink pens had been checked!) where the intensity of infrared light measured by our 'thermometers' was being recorded, and at 10 h 58, I made the happy announcement: "We've seen the dust [of the rings], we've seen it." Under Alain's vigilant eye, Don and I would roll out our programme during this totality that would exceed 1 h on the clock. Now sure that we had detected the much hoped for infrared signal from the ring of dust, we began to analyse it with a spectrograph to determine the composition of the microscopic grains making up the ring. Under the stars at high noon with all its navigation lights on, Concorde hurtled on due east into the night. We renewed the magnetic tape of the register. Pierre announced over the radio: "Third contact in six minutes." The last few scans, almost a routine soon to come to an end. And 3 min and 25 s before this third contact, we had ended our measurements and dashed to the side portholes of the cabin to marvel at the extraordinary spectacle and finally get our own photos. I am almost ashamed to say here that we thus neglected our measurements, even though completed, for 3 min stolen from the totality, a length of time for which most astronomers are ready to mount expeditions to the other side of the world!

A part of Africa had been plunged into darkness by the huge shadow cast upon it by the Moon, within which we were still flying. Beyond the edge of the shadow, toward the south, the savanah, yellow with drought, stretched away into the penumbra, as far as the eye could see, where distant towers of cumulonimbus threatened storms in the intertropical convergence zone above Niger. In the foreground, the white gothic wing of the aircraft reflected a glowing horizon, rounded by the curvature of the Earth. Above the horizon, far away and hence lit up by the Sun, the Earth's atmosphere scattered light, milky white in the lower layers of the troposphere where sandstorms had whipped up dust, then darker and darker blue as one looked higher into the stratosphere, until one became lost in the darkness of night (Fig. 5.4). Jean, too, was taking photos of this extraordinary scene, and one of them would soon come to the attention of the tabloids.

After the third contact at 12 h 07 min 13 s, the lunar disk once again revealed the reddish glow of the chromosphere to John. We were flying at 17,600 m because the plane, lighter now that it had consumed a good part of its kerosene, had been climbing ever higher into the stratosphere. Pierre passed me a paper that I still keep carefully after all this time. It confirmed that we had enjoyed 72 min of totality, almost exactly ten times the maximal duration of an eclipse when it is observed at ground level. We had just beaten the record, observing totality for longer than the cumulation of all eclipses viewed for a quarter of a century. As Pierre noted later on: "Those who had christened this the longest eclipse of the century didn't realise just how right they were".

Throughout the flight, Hubert Guyonnet, the radio navigator, answered questions raised by the pilots: "Our ground speed?–1102 knots." (This would be 565 m per second, or 2037 km an hour.) "Predicted time of arrival at 23°36′N?–10 h27." "Distance to Niamey?–480 nautical miles." "Wind?–46 knots from 087 [an east wind]." Those were the 46 knots of opposing wind that had prevented us from reaching the hoped for 80 min of totality. From Jean Dabos, the copilot, André Turcat was asking mainly for weather reports at Niamey, Tamanrasset, and Kano, which were alternative landing sites in case of rerouting, until they were no longer

**Fig. 5.4** During totality on 30 June 1973, an extraordinary spectacle was observed through a side porthole of Concorde 001 at an altitude of 17,000 m. In the foreground, the 'gothic' delta wing of the plane shows a faint reflection of the horizon. In the background, the glowing horizon reveals the curvature of the Earth. To the south, on the African landscape yellowed by drought, there is a clear boundary between the umbra and the penumbra. Far away, the lower atmosphere is milky white, scattering the Sun's light. Higher up, above an altitude of 12 km, the stratosphere scatters a beautiful blue light. © P. Léna

attainable, in order to be able to make decisions soon enough on the basis of progress information provided by Perrier and Guyonnet. But in the end the weather was fine everywhere, perhaps too fine for all these thirsty countries, and the threat of the intertropical weather front kept well away from us (Fig. 5.5).

The light from the solar disk was returning little by little, and 12 min after third contact, our colleagues' final observations were completed and the plane went into a bend toward the south. Forty years later, our navigation mechanic Michel Rétif wrote to me in the following words[5]:

> The engine regime was being reduced when suddenly one of them, instead of stabilising and slowing down, actually stopped altogether, without any alarm going off. In this situation, in supersonic flight, the non-functioning engine begins to rotate on its own due to the air flow, and this has the effect of maintaining the electricity supply to the aircraft. In subsonic flight, the consequences of such an incident would have been spectacular, with the corresponding loss of electricity generation. But this engine restarted very quickly and subsequently accepted the slowdown.

[5] Michel Rétif: Personal communication, 2013.

**Fig. 5.5** On this stamp, issued by the Republic of Congo in 2009 to commemorate the International Year of Astronomy, Concorde is still remembered, as the commercial airliner wearing French colours flies before the terrifying cumulus clouds that characterise the intertropical convergence zone. Documents kindly provided by Henri Aubry

We locked our equipment in place and went back to the landing seats, ready to touch down on the runway at Fort Lamy in Chad some 49 min later. To check that the runway was indeed clear, and despite the fact that there was not much fuel left in the reservoirs, André Turcat made a first low altitude transit over the site. The weather was magnificent, the Sun still half eclipsed, and a kind of euphoria came over me as the wheels of the Concorde touched the ground in Chad. The exit steps were put in place and we climbed down, still wearing our red suits. In a flash, I thought of my father who, now aged 81 and always enthusiastic about exploration of Africa, would be eagerly awaiting a telegram in Paris to announce our success.

The Caravelle carrying all our other colleagues was much slower and would only arrive at Fort Lamy during the afternoon, long after us.

## A Visit to the Nyangatom People of Ethiopia

Travelling at more than 2000 km an hour, the Moon's shadow whose pursuit we had just abandoned continued on its way to skim past Ethiopia and the southernmost tip of Somalia, before disappearing across the Indian Ocean.

The Nyangatom are an extremely isolated people living in the region between southern Sudan and Ethiopia. Two French ethnologists, Serge and Marie-Martine Tornay, paid them a visit. They warned them about the eclipse that was about to occur. Every month a soothsayer anoints the villagers with white clay, sprinkling them with saliva and chewed up aloe leaves, in order to bring back the Moon at the time of the new Moon. Let us turn to the ethnologist,[6] but using Universal Time rather than local time for easier comparison with the story we have told so far:

[6] The quote comes from a research note, published by Gérard Francillon in 1974. The full analysis of the ethnological observations made in Ethiopia is given by Serge Tornay: *Médecine du corps, médecine du cosmos: l'éclipse chez les Nyangatom*. In: *Soleil est mort: l'éclipse totale du 30 juin 1973*, G. Francillon and P. Menget (eds.), pp. 201–243 (1979). See the bibliography.

**Fig. 5.6** On 30 June 1973, the lunar shadow brushed past the southernmost point of the Federal Republic of Somalia as it left Africa. This unfortunate country, affected by a devastating drought and governed by a remorseless dictatorship, would gradually sink into tragedy. Documents kindly provided by Henri Aubry

"13 h11. The soothsayer continues the sprinkling. He spits upon all the women in the hamlet who turn up. The eclipse seems to be at maximum. A woman is heard to shout: 'Darkness, all is darkness!' Only then do the others notice that something unusual is happening. The soothsayer examines the Sun and confirms that it is partially dead.

"13 h15. The light is gradually returning. Suddenly someone shouts: 'Strike the gourds, strike the gourds to bring back the Sun. Hit them hard, it's coming back, it's like the Moon.' The women then begin the necessary pounding and pummelling.

"13 h21. The noise dies down. We return to the soothsayer. 'Did you see?' he asks. 'It has returned and it seems contented.'

"14 h16. The country has recovered."

"The cultural background of these inhabitants of Ethiopia provides them with a ritual response to the event. For them, the cosmic phenomenon is viewed as a sign, rather than as an event."

So concluded the ethnologist Gérard Francillon,[7] while the shadow went on its way (Figs. 5.6 and 5.7).

[7] Gérard Francillon in *Note de recherche: Eclipse de soleil du 30 juin 1973* (1974), unpublished. The publication *Soleil est mort* (op. cit.) contains a much more detailed analysis.

**Fig. 5.7** In the lower Omo valley in Ethiopia, 5 min after totality during the eclipse of 30 June, this Nyangatom woman strikes an upside-down milking pot. Photographed by the ethnologist Serge Tornay, she turns her back to the Sun. © Labethno, Société d'ethnologie, 1979/S. Tornay. See note 39

## Mach 2.05 or above?

Almost 40 years later, in a moving letter sent to me just before he passed away, Henri Perrier alluded to his indictment after the tragic accident in Gonesse, before the final judgement in which the charges against him were dropped. He also gave me his own version of our race against the Moon's shadow, which I record here[8]:

> Among my recollections of these rehearsal flights, and then the flight itself, two of them remain vivid: the accuracy in the specification of the time to within one tenth of a second [the click of the clock], while the best estimate of our geographical position using the inertial guidance systems was between 1 and 1.5 nautical miles (i.e., equivalent [in time] to

[8] Henri Perrier: Personal communication (2011).

| TEMPS | ZP (M) | MACH | TA (DK) | TETA (DEG) | PHI (DEG) | VSOL (M/S) |
|---|---|---|---|---|---|---|
| 11-35-10.00 | 17356. | 2.10 | 180.1 | 64.4 | 0.2 | -551. |
| 11-35-11.00 | 17356. | 2.09 | 180.5 | 64.4 | 0.2 | -551. |
| 11-35-12.00 | 17357. | 2.09 | 180.9 | 64.4 | -0.1 | -551. |
| 11-35-13.00 | 17357. | 2.09 | 180.9 | 64.3 | -0.1 | -551. |
| 11-35-14.00 | 17356. | 2.09 | 180.3 | 64.4 | -0.1 | -551. |
| 11-35-15.00 | 17356. | 2.10 | 180.0 | 64.4 | -0.1 | -551. |
| 11-35-16.00 | 17356. | 2.10 | 179.9 | 64.3 | -0.1 | -551. |
| 11-35-17.00 | 17356. | 2.10 | 180.1 | 64.4 | -0.1 | -551. |
| 11-35-18.00 | 17356. | 2.10 | 179.9 | 64.3 | -0.1 | -551. |
| 11-35-19.00 | 17356. | 2.10 | 179.9 | 64.4 | -0.1 | -551. |
| 11-35-20.00 | 17355. | 2.10 | 179.7 | 64.4 | -0.1 | -551. |
| 11-35-21.00 | 17355. | 2.10 | 179.7 | 64.2 | -0.1 | -551. |
| 11-35-22.00 | 17355. | 2.10 | 179.7 | 64.4 | -0.1 | -551. |
| 11-35-23.00 | 17355. | 2.10 | 179.7 | 64.4 | 0.2 | -551. |
| 11-35-24.00 | 17356. | 2.10 | 179.8 | 64.2 | -0.1 | -551. |
| 11-35-25.00 | 17356. | 2.10 | 179.9 | 64.2 | 0.2 | -551. |
| 11-35-26.00 | 17357. | 2.09 | 180.6 | 64.4 | 0.0 | -551. |
| 11-35-27.00 | 17357. | 2.09 | 180.7 | 64.3 | -0.1 | -551. |
| 11-35-28.00 | 17357. | 2.09 | 180.7 | 64.4 | -0.1 | -551. |
| 11-35-29.00 | 17357. | 2.09 | 180.9 | 64.4 | 0.2 | -551. |
| 11-35-30.00 | 17357. | 2.09 | 180.9 | 64.4 | 0.2 | -551. |

**Fig. 5.8** Extract from the onboard recording of 30 June, used to determine the exact trajectory of Concorde 001. It is 11 h 35 min 10 s and the plane is flying at 17,356 m at a speed of Mach 2.09–2.10. As the Mach number is calculated from the local air temperature at this altitude, this number varies, even though the ground speed shown in the last column remains constant. © P. Léna

between three and five seconds); and the fact that, in order to compensate for a wind component along the flight path that was greater than had been registered during the flight on 28 June, Michel Rétif and I, with the unspoken agreement of Turcat and no longer restricted by the external observations, had decided to cruise at Mach 2.05 by removing the [automatic] control systems which reduced the thrust as soon as we went above Mach 2.01 to 2.02. To do this, on the chart recorder at my post, I could observe the significant pressures of the surge margins and use these to tell Rétif what manual display [to choose] in the position of the air inlet ramps. This was a 'reasonable' operation as long as our trajectory was not expected to subject us to sudden changes in static temperature, according to our rather vague knowledge of the weather in this part of the atmosphere and at these latitudes. Having spoken of this with my old boss [Turcat], I know that he had no recollection of the risk taken here with regard to a compressor stall, which would have been impossible to make good on the prototype without significant deceleration, and this to gain perhaps one minute of observation.

I have kept the recordings of the flight parameters just as they were stored by the aircraft test facility. I would read the Mach number minute by minute during the flight. And there was no doubt: we went up to Mach 2.09 for several minutes! According to this same recording, our ground speed at this precise time was 551 m per second, or 1983 km an hour (Fig. 5.8).

Did we actually fly faster than the shadow? That would be another record, although of a more sentimental nature. The shadow never moved more slowly than 601 m per second, and our average ground speed throughout totality was 568 m per second, but with variations that no doubt took us very close to 601 m per second. If only the wind had not been against us!

## A Failed *Coup d'Etat*

Fort Lamy appeared to be filled with scenes of jubilation. We were coming down the main avenue with André Turcat in the lead, surrounded by crowds of people, many of them carrying blackened glasses to watch the eclipse, as totality had ended more than an hour before. With some surprise, we noted machine gun posts set up at certain crossroads. The explanation was soon made clear by Mr. Baldit, the representative from the French embassy who had come to greet us. There had just been an aborted *coup d'état*, but we did not know whether its architects, castigated by the article quoted below, had deliberately chosen to make it coincide with the eclipse.

Two days later, on 2 July, the Chad press agency published the following slightly pompous account in its roneoed bulletin which I still keep, under the title *Le Canard déchainé et les conspirateurs*[9]:

> Chad has been living through dreadful times since achieving independence [...]. If the Chadian people have been able to stand up to this trial, it is thanks to the composure, and above all the unbending patriotism of François Tombalbaye [the president]. Just a little further and Dopélé would have plunged us last week into the midst of the most catastrophic events, and by now Chad would be a blazing inferno or a gigantic field of dead bodies.

So we had a narrow escape! This same issue of the newspaper tells of the arrival of Concorde and quotes André Turcat's remarks to the journalists:

> We know the tremendous problems you face [the drought], but the task of understanding the world does not need to be socially motivated in order to justify itself. Research has consequences for everyone, and what appears to be pure science is never far from applications for the benefit of all.

Words that remain as relevant today as ever (Figs. 5.9 and 5.10).

**Fig. 5.9** André Turcat (*left*) after the landing at Fort Lamy. Facing the camera, one of the Aérospatiale engineers. Behind him, Donald Hall wearing a red suit. © P. Léna

[9] Chad Press Agency, Daily Bulletin: *InfoTchad*, no. 2809, 2 July 1973 (roneoed). The title is a play on words, *Le Canard enchainé* being a well known satirical newspaper in France (the word '*canard*' is slang for a newspaper).

**Fig. 5.10** Reception in the town of Fort Lamy. In the foreground, *left to right*, the scientists Yves Viala, Jean Bégot, and Michel Ravaut, accompanied by a small crowd of onlookers. The eclipse had just ended. © P. Léna

That evening, when we arrived somewhat exhausted at the hotel, we examined our results, and Pierre Charvin sent the crucial telegram to the International Astronomical Union in which he collated all the astronomical events and registered our record. We also contacted our research institutes, which would have to answer questions posed by journalists. At first glance, all the instruments seemed to have worked well, we had harvested a considerable amount of data on the corona, and we were fully satisfied with the way things had gone. So many unexpected incidents might have brought failure to our mission: we might have run out of cryogenic fluids, poor weather might have forced a reroute, the recording devices might have broken down, a porthole might have burst, vibrations might have gone out of control, or there might have been a problem with an engine air inlet.

## The Return Home

Two days later, the 001 prototype left N'Djamena for Toulouse, stopping over for 6 h in Algiers, where we took members of the government aboard. The correspondent of *Le Monde* made the following observation: "The Algerian authorities were visibly sensitive to this mark of respect." Its duty accomplished, the Caravelle had made a stopover on the island of Djerba in Tunisia, and the whole team had bathed in the mild waters of the Mediterranean. After dismantling our observation instruments, not without some sadness, each of us left Toulouse, taking our measurement results with us for analysis. Apart from myself and my friend Don Hall. On the evening of our arrival, Don was struck by a terrible African fever and had to be hospitalised in Purpan. In his room, I watched over him with great concern, but the

**Fig. 5.11** At the Air and Space museum in Le Bourget (Seine-Saint-Denis, France), 40 years on, the Concorde 001 prototype still carries the eclipse logo from 1973 on the side of the fuselage. © Musée de l'Air et de l'Espace - Le Bourget/Alexandre Fernandes

robust constitution of his 28 years, celebrated only the day before the eclipse, soon had him out of danger, although ten kilos lighter.

After its last flight on 19 October, the Concorde 001 prototype, under the matriculation F-WTSS, came to land at Le Bourget, having accomplished a grand tour over the Atlantic (Fig. 5.11). On board was an 82 year old lady, the Hungarian countess Lilly von Coudenhove-Kalergi,[10] who was one of the first women to obtain a pilot's licence, back in 1912. She was so enthusiastic about this plane that Aérospatiale offered her the 397th and last flight made by the prototype. Concorde was handed over to General Lissarague who was the director of the brand new *Musée de l'air et de l'espace*. Formally and symbolically, André Turcat was to present the keys of the plane to the general. But unfortunately there is no key for a test plane, so caught short, and not wishing to disappoint, Turcat took his car keys out of his pocket! Concorde 001 had flown in supersonic flight for a total of 812 h 09 min, of which 254 h 45 min in supersonic flight. André Turcat, who was the first and the last to fly this extraordinary prototype, made the following address[11]:

> "It made too much noise and too much smoke. It didn't have sufficient range to cross the North Atlantic. It was jam packed with test equipment. It didn't have that watch with 24 hour dial which doesn't tell anyone the time, but which is required by certain airlines. And to be quite honest, the ash trays were all full. The only thing left for this old chap, already five years old, is clearly the museum. It achieved quite a lot though. It was the first to carry 140 tonnes in supersonic flight. It flew from 200 to 2300 km/h. It crossed the South Atlantic in two hours and went beyond the 45th parallel south. It climbed, nose-dived, and

[10] http://www.pionnair-ge.com/spip1/spip.php?article441

[11] A. Turcat: *Concorde: essais et batailles*, pp. 265–266. See the bibliography.

> rolled, and it was pushed and mistreated as much as any other plane. It underwent thousands of modifications. It was a docile subject in the hands of ten test pilots and twenty airline pilots. It was admired for its power and elegance, now surpassed by its successors. It carried the most famous of people. It spent more than an hour under the total eclipse of the Sun, a record that will take a very long time to beat. It measured the nitric acid in the stratosphere and watched the shadow of the Earth on the depth of the sky. It gave us the greatest joy of our careers, and it announced a new era in world transport.
>
> "In a word, we simply adore it."

Concorde 001 is still at the museum in Le Bourget, joined by one of the commercial airliners of the Air France series, F-BVFF. At the very front of its fuselage features the logo of the African eclipse and, on the ceiling of the cabin along which thousands of visitors now walk, the four portholes are still there.

# Chapter 6
# Seventy-Four Minutes of Observation, but What Gain for Science?

It's all very well setting up a record observation time, but that wasn't really the main objective for the scientists who set off on this adventure. One might say that it was only a secondary result. It was the Sun and its corona that really interested us. So what discoveries were made thanks to the three exceptional circumstances we benefited from: the long period of totality, the slow scan of the chromosphere just before and just after totality, and the excellent access to infrared emissions from such a high flight path?

## In Which the Sun May Lose Its Rings

So are there rings or shells of interplanetary dust concentrated around the Sun in the plane of the ecliptic, that is, in the plane containing, to a good approximation, the orbits of all the planets? Our predecessors made this assertion on the basis of rather difficult ground-based observations during the eclipse of 1966. After carefully analysing our observations of the infrared light from the corona, Don Hall and I reasserted this claim in a talk based on our results and presented in São Paulo at the end of 1973. However, as often happens in science, things turned out to be much more complex than we had assumed in those days of innocence, and the subject led to much debate.

When we observe a sunset at low latitudes in a pure moonless sky, we can make out on the western horizon a pale and delicate glow extending upward almost vertically, well above the Sun as it slips down over the horizon (see Fig. B.1 in the appendices). This is the so-called zodiacal light. It is due to the scattering of the Sun's light by grains of dust concentrated in the plane of motion of the planets, known as the ecliptic plane. This in itself shows that the grains are present well beyond the immediate suburbs of the Sun where we have been focusing on them up to now. Apart from this clue, there are many reliable demonstrations that these interplanetary grains really exist, constituting one of the ingredients of the solar



P. Léna, *Racing the Moon's Shadow with Concorde 001*, Astronomers' Universe,
DOI 10.1007/978-3-319-21729-1_6

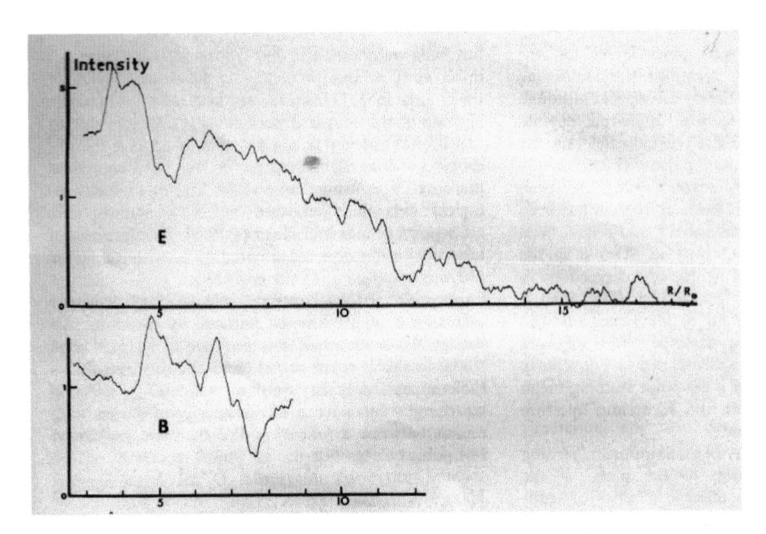


**Fig. 6.1** Original recording of infrared emission by dust shells, as we considered we had detected them during totality in the flight of 30 June. Each of the two curves shows a measurement of the infrared radiation intensity from the F corona. These two measurements were made during totality, moving away from the centre of the Sun along the ecliptic, through a distance of between about four and fifteen apparent solar radii (the apparent solar radius seen from Earth is 0.25°). Credit: Léna et al., A&A, 37, 81-86, 1974, reproduced with permission © ESO

corona, the one denoted by the letter F. Either they belong to the primitive cloud from which our Solar System originally formed, or they result from evaporation of cometary nuclei whenever comets graze close past the Sun during their periodic returns to the inner Solar System (Fig. 6.1).

For their part, physicists calculate the behaviour of a tiny rocky grain of interplanetary material—measuring just a fraction of a thousandth of a millimetre—when it is subjected to the Sun's gravity, which attracts it, whilst receiving light which tends to push it away from the Sun, heating it, and in some cases causing it to vaporise. Furthermore, various kinds of crystal grains, well known by dust gathered from shooting stars in our atmosphere or even in space, will display quite different behaviour depending on whether they are made from iron-bearing or magnesium bearing olivine, for example, to cite only this particular mineral, a silicate. So the question here is not whether such grains transit in the immediate neighbourhood of the Sun, before they are heated up to more than 1500 °C and evaporate, but whether these grains can concentrate sufficiently to form rings or shells.

Forty years on, as I write about this true story of our eclipse, and following a great many observations made from high mountain tops, such as the Hawaii observatory at 4200 m altitude,[1] or from a Japanese stratospheric sounding balloon, the results remain contradictory across the vast numbers of published papers, despite the fact that each piece of research is of the highest quality. Some also

[1] P. Lamy et al.: No evidence of a circumsolar dust ring from the infrared observations of the 1991 solar eclipse. In: Science **257**, 1377 (1992).

measured infrared emission from the corona, at a distance of a few solar radii from the centre of the star, while others, just as respectable and conscientious, observed nothing of the kind. This already shows that, perhaps with a certain naivety, we didn't choose an easy problem when we selected such a complex measurement programme for our Concorde flight. From these apparently contradictory results, it now seems possible to draw two quite different conclusions: either all those who, like ourselves, measured coronal infrared emission suggesting the existence of a dust ring were completely mistaken, for example, because their instruments were not sensitive enough, or suffered from unidentified background signals, or they misinterpreted those measurements; or else the rings do exist but not all the time, something which might reconcile the two series of observations.

Clearly, the second conclusion appeals much more to me than the first, even though, as a scientist, one must always be ready to admit error. From the 1980s, and for many reasons, my work drifted away from the corona and its dust and I ceased to work on the problem. But in 2010, Lyubov Shestakova, an astronomer at the Fesenkov Astrophysical Institute in Kazakhstan, published an article that attempted to reconcile the two possibilities, suggesting the possibly sporadic existence of circumsolar rings.[2] But other specialists consider all doubt here to have been removed. According to Shestakova, this sporadic dust production could explain why, at certain times, observations reveal the presence of rings, until they sublimate and dissipate. No further observation would then reveal their presence, until new rings are formed. I like to think that future observations will come down in favour of this solution, but whatever happens, in 1973 we were able to make our contribution. Such is the way of science!

## In the Vicinity of Beta Pictoris

What did the Kazakh scientist base her reasoning on? Here, there is a beautiful story of astronomy that I cannot resist sharing with my readers here. Eleven years after our Concorde flight, in 1984, at a time when infrared light from the heavenly bodies had begun to interest the whole scientific community, two astronomers in the United States, Bradford Smith and Richard Terrile, analysed images that had been obtained by the first infrared light observatory sent into space by NASA a few months previously.[3] Apart from its telescope, this satellite takes the form of an enormous Thermos flask equipped with 'thermometers' which have progressed enormously in sensitivity. Around a particular star denoted by the Greek letter 'beta' in the southern hemisphere constellation known as Pictor, the Painter, Smith and his colleague detected a vast disk of dust orbiting around the star. A superb

[2] L.I. Shestakova et al.: The velocity of the dust near the Sun during the solar eclipse of March 29, 2006 and sun-grazing comets. In: 2010arXiv1003.2818S (2010).

[3] Infrared Astronomical Satellite (IRAS), launched by NASA on 25 January 1983.

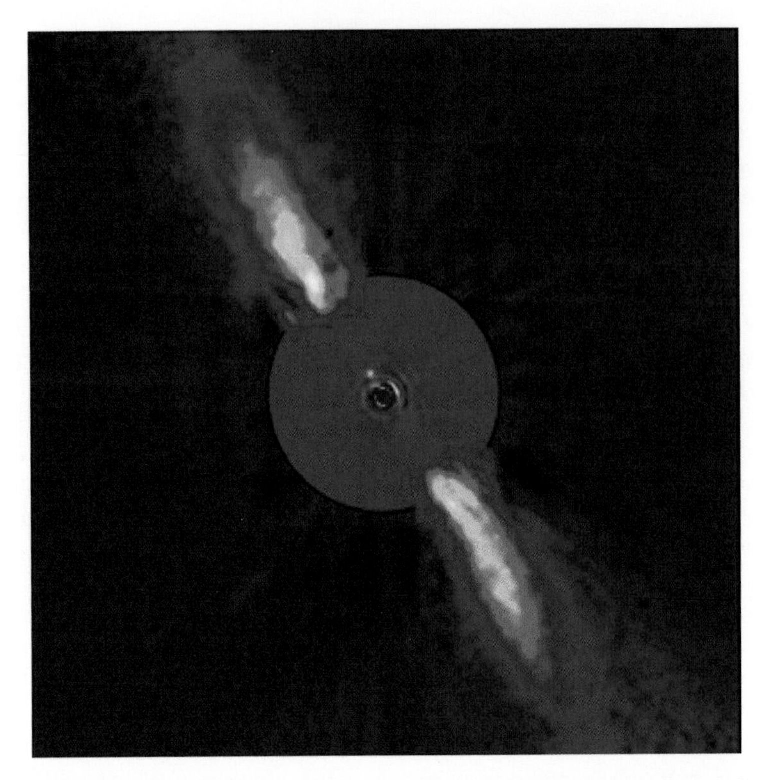

**Fig. 6.2** Beautiful image of the disk around β Pictoris (or Beta Pic), observed in the infrared by one of the telescopes at the European Southern Observatory (ESO) in Chile. In the centre, the star itself is hidden by a coronagraph mask, producing a kind of artificial eclipse. The disk around the star is viewed edge-on and extends out a long way. It is artificially attenuated in the inner region to reveal the planet β Pic b as a point of light, observed in 2008 by the Very Large Telescope (VLT). A few years later, the planet was observed again on the other side of the star, and its orbital motion is still being monitored. © A.-M. Lagrange/CNRS

image of this disk was subsequently obtained by an Earth-based telescope. And until today, this 'Beta Pic' as it is known affectionately to astronomers, has been the cause of great fascination. For this young star and its disk seem to provide all the conditions needed for the formation of planets, precisely by agglomeration of microscopic grains in its disk. Indeed, the close similarity with the grains in the solar corona was immediately noted, to the point where disks like this were qualified as exozodiacal, with reference to the solar zodiacal cloud. So are we not looking here at a sort of replica of what may originally have been our own Solar System some 5 billion years ago, at the time when the youthful Sun was still surrounded by its disk of dust and gas, just before its planetary escort came into being? (Fig. 6.2)

As soon as Smith and Terrile published their discovery, a group in the same laboratory as Serge, the Paris Institute of Astrophysics (IAP), set to work to monitor Beta Pic on a regular basis. It was not long before the group leader, Alfred Vidal-

**Fig. 6.3** Photographed in April 2004 by the Large Angle and Spectrometric Coronagraph (LASCO), a spaceborne instrument carried aboard the Solar and Heliospheric Observatory (SOHO), Comet Bradfield approaches the Sun (hidden by a mask here, the size of the solar disk is shown by the *white circle*), releasing gas and dust produced by evaporation of its nucleus into a long tail. © ESA/NASA

Madjar, a colleague who had vigorously supported me when we were preparing our eclipse instrument, discovered that the light from the star underwent sudden and surprising variations. The explanation put forward by Alfred, now largely confirmed, is as follows. Comets form from the gas and dust in the disk, similar to the hundreds of thousands of comets that populate our own Solar System. Attracted by the star's gravity, they pass close by and evaporate in the intense radiation which heats their nucleus to produce a magnificent cometary tail, or even fall right into the star itself. The Kazakh astronomer, Lyubov Shestakova, mentioned earlier thus stressed the relationship between the dust in our own Solar System and this Beta Pic phenomenon (Fig. 6.3).[4]

Beta Pic has led to other revelations, since in 2008 Anne-Marie Lagrange, one of Alfred's students who has now become a prominent astrophysicist, discovered a

[4] L.I. Shestakova: The beta Pictoris phenomenon near the Sun. In: Astronomical and Astrophysical Transactions **22** (2), 191–211 (2003), 10.1080/1055679031000080339

planet in the disk. She published an image and has been monitoring its motion around the star ever since. But that's another story.

## How to Heat a Gas to a Million Degrees

We have already encountered the modest and smiling Donald Liebenberg, a veteran airborne eclipse observer. He built a beautiful instrument for Concorde, using a TV camera to film the corona. At the time, a more fundamental question than the nature of the rings bothered solar specialists. The gaseous hydrogen at the surface of the Sun, which is what we see in the sky and which sends us its beautiful white light, is not really at a very high temperature, in fact, barely 5000 °C, nothing compared to the millions of degrees in the Sun's core. Above this surface, the temperature decreases slightly—it was this decrease that I was studying aboard the NASA Galileo I aircraft–, but then it increases suddenly, and the hydrogen in this region a few 1000 km higher is already at 20,000 °C. The lunar limb sweeps past this thin layer, the chromosphere, in a brief instant when viewed from the ground. It is the light from this layer that gives rise to the so-called red flash, which ancient myths interpreted as the blood of the Sun when it was eaten by the Moon, just before and just after totality. Barely 1000 km higher, the gases begin to rarify and this is the corona proper, where the temperature goes on increasing until it very quickly reaches a million degrees. This sudden increase in temperature in the chromosphere and the corona is not so easy to explain: light from the Sun cannot heat this gas because it is much too rarified to absorb it, so one would expect the temperature to fall off with altitude. But what then could be the cause of this heating? This was, dare we say it, the burning question which preoccupied Donald as much as Serge or John when it suddenly became possible to carry out a lengthy observation of this region from Concorde.

If light could not heat the gas, perhaps some other waves could do so. But what kind of waves? When we speak of waves, we think of oscillations, time varying quantities, like the vibration of a violin string or a wave on the surface of the sea. Now it was precisely the slowing down of the Moon's motion relative to the Sun due to our supersonic flight that gave time for time to pass, either at second contact when the Moon begins to cover the chromosphere, or at third contact when it reveals it again, or during totality, when there would be plenty of time to observe any periodic light variations that such oscillations would be sure to produce. In fact it has been known since the 1950s that sound waves, also known as acoustic waves, are permanently present in the lower layers of the Sun, and their period of oscillation has been established as 5 min. This was a major discovery which, even half a century later, still inspires study of the Sun, and today also the stars. After his 1955 flight in Indochina, Raymond Michard, mentioned earlier as president of the Paris Observatory in 1972, was one of those who made this discovery during a visit to the United States. Could it be that these acoustic waves, born in the depths of the Sun, propagate toward the surface where they dissipate their energy in the form of heat and thereby raise the temperature of the coronal gas? It was an attractive hypothesis that deserved to be put to the test at supersonic speeds!

The search for acoustic waves was clearly the programme of Donald Liebenberg and also Serge Koutchmy, a programme carried out under his porthole by Jean Bégot. For his part, John Beckman wanted to determine more precisely whether the 20,000 °C temperature of the chromosphere was uniform, or whether there was a mixture of cold regions coming up from deep down and referred to as spicules. To measure the tiny infrared radiation from the chromosphere, John was not afraid to face the difficulties of using a cooled 'thermometer', like our own. Meanwhile, Donald and Serge were observing, taking the spectrum, and photographing the infrared light of the chromosphere and the white light of the corona using robust and well understood techniques. This is how science works, combining well tested methods with more audacious ones.

From the resulting measurements, Donald identified vibrations in the coronal gas, showing that it too oscillates with a period of 5 min, while the 200 and 50 frames obtained by Jean and analysed by Serge, who had also made successful observations of the eclipse from Moussoro in Chad, would eventually contribute to eliminating the hypothesis that acoustic waves were responsible for coronal heating. Our two British colleagues, John Beckman and Paul Wraight, also published very good results. In the end, it was a good harvest of data for science.

## Was Concorde Spied Upon by a Flying Saucer?

It took the immensely tragic destruction of the twin towers of the World Trade Center in New York on 11 September 2001 to show the citizens of the United States that their country was not some kind of impregnable fortress that no one would ever be able to penetrate. No one, that is, except extraterrestrials! This fear comes from along ago[5]:

> On 30 October 1938, the CBS radio station in New York announced the live broadcast of a variety show given in one of the grand hotels. Five minutes later, the music was stopped to give a news flash, according to which the Mount Jennings observatory in Illinois had just observed curious and inexplicable vapour projections on the surface of the planet Mars. Then seismographs in New York recorded a tremor. A few minutes later, the somewhat shaky voice of the speaker announced that a large and mysterious object had just crashed into a factory in Princeton, New Jersey, and burst into flames. There were sounds of sirens screaming as the first emergency services arrived on the scene. Then, finally, a reporter began to give a 'live' commentary on the terrifying scenes taking place before his eyes. The Martians were disintegrating anyone who came near the object . . .

The author of this radio hoax, which actually triggered scenes of panic in New York, was none other than a young actor by the name of Orson Welles, then

[5] http://www.forum-ovni-ufologie.com/t8611-la-guerre-des-mondes-les-martiens-d-orson-welles-le-30-octobre-1938#ixzz2a3iZLX5O

only 23 years old, who made himself famous in the process. His aim had been to make a radio adaptation of *The War of the Worlds* by his close namesake, the writer H. G. Wells.

After the end of the war in 1945, the Cold War stepped in to whip up a siege mentality in the United States.[6] The newspapers were full of reports of flying saucers, and even sometimes their crew. The US Air Force set up Project Blue Book, which ran from 1947 to 1969. During the McCarthy era, in 1952–1953, the US counter-intelligence agency (CIA) carried out its own inquiry, in great secrecy, of course. The US Congress was concerned about radio or any other signals that might tell potential invaders about the presence of inhabitants on Earth. France was not spared this strange fascination, with tabloid newspapers like *Paris Match* rushing out to report the slightest gleam of suspicious light in the sky. The government could not ignore public concern over this issue and the National space research institute (*Centre national d'études spatiales*, CNES) set up a special department to investigate unidentified objects spotted in the sky (*Groupe d'études et d'informations sur les phénomènes aérospatiaux non identifiés*, Geipan),[7] which was still running in 2014. One particular astronomer, belonging to the same research institute as Serge but a bit of an outsider, became the champion for the extraterrestrial origins of UFOs (unidentified flying objects), and the press made great use of his testimony to back up some of their less reliable claims.

Now, in one of the 36 Kodachrome slides of the African horizon taken by Jean at 12 h15 with his own 24 × 36 format camera, during the few brief moments when he had been able to contemplate the spectacle described above through a side porthole of 001, a suspicious dash of light appeared just above the horizon, against the perfectly black background of the stratosphere (Fig. 6.4). After enlarging the snapshot, this light became a tiny cloud with an apparent size close to that of the full Moon, white in the middle, orange-red toward the top, and fringed with green (Fig. 6.5). When the film was checked, it was shown that this was not due to a defect or scratch. Through an understandable lack of caution in Serge's lab, the press was informed in the days following our return and published the photo. There followed much speculation. Was it the Soviets, whose Tupolev Tu-144 had so sadly crashed during a demonstration flight at the Paris Air Show in Le Bourget only a month earlier, who were spying on Concorde as it raced after the Moon's shadow? Or was it extraterrestrials, fascinated by this machine, a potential competitor for their own, who had also organised a rendezvous with a cosmic phenomenon that could be spotted from anywhere in the Solar System? And so on and so forth. What were these scientists hiding from us, given their lack of interest, even contempt, for UFOs?

The National research agency (CNRS) was put in a difficult position and Serge Koutchmy had to give an explanation. To do this, he took his time and deployed all his many talents. In November 1973, he sent a brief statement to the *Agence*

[6] https://en.wikipedia.org/wiki/Unidentified_flying_object

[7] https://www.cnes-geipan.fr

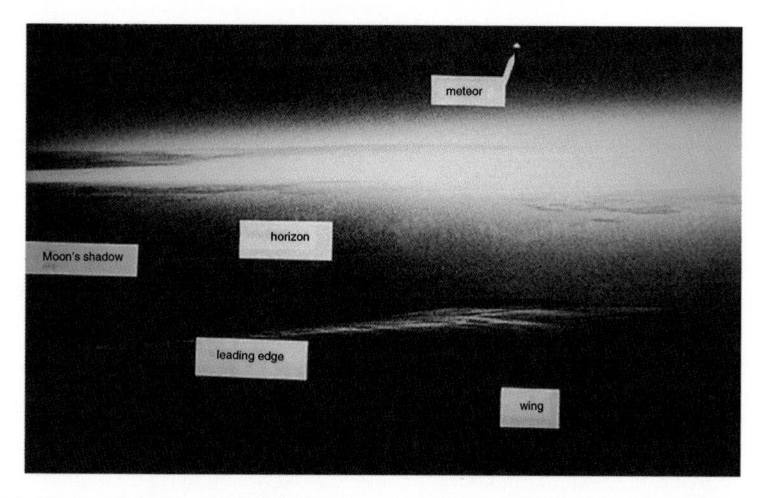


**Fig. 6.4** Black and white photo taken from Concorde 001, showing a mysterious point of light. © J. Bégot & S. Koutchmy IAP CNRS

**Fig. 6.5** Enlarged view of the object in Fig. 6.3. This image was used to prove that it was in fact a meteor. © J. Bégot & S. Koutchmy IAP CNRS

*France-Presse*, a Paris-based international news agency, detailed in a confidential and strictly scientific note, dated 8 April 1974 and passed on to the authorities. This is how it read:

> The physical and geometric features of the cloud show that the most likely hypothesis to explain this observation is a meteor shower in the upper atmosphere. This conjecture was made soon after the photo was published (statement to the *Agence France-Press* on 4/11/1973). The following elements support this identification: (a) size and distance of the cloud,

> determined using geometric, photographic, and optical parameters; morphology of the appendix [of the spot of light]; (b) colour suggesting emission of sodium D lines; (c) maximum visibility of the Beta Taurid [meteor] shower during totality [of the eclipse].

This UFO was therefore nothing other than an ordinary meteor of a few grams, burning up as it shot through the stratosphere at an altitude of some 50 km. Its green colour, similar to the colour of the aurora borealis, could be attributed to oxygen atoms present at this altitude, excited by their interaction with the passing object. It turned out that, at the date and time of the eclipse, the Earth happened to intersect the trajectory of some well known meteorites as it went on its orbit around the Sun. Too bad for cheap sensationalism and so much the better for the scientific rigour of Jean Bégot and Serge Koutchmy!

# Chapter 7
# The Future of the Solar Eclipse

Human beings are unlikely ever to be indifferent to the magnificent display of the Sun suddenly blacked out and crowned in glory, stars shining at midday, the hush that falls swiftly over nature, mysterious shadows sweeping across the ground, and the blood red flash of the solar chromosphere lasting only for a split second. A show that will go on repeating itself for as long as the Moon has not moved too far away from the Earth, a distant time when the Moon's disk will have shrunk too much to be able to blot out the Sun.

The rare and spectacular sight of a total eclipse of the Sun will always be worth a long journey, and scientific tours, even airborne ones, are becoming popular nowadays, for amateur astronomers as well as laypersons. Someone who organizes such trips, Xavier,[1] already mentioned in this book, generously invited me to watch the total eclipse of 20 March 2015, somewhere above the Atlantic Ocean between Iceland and Scotland, aboard a magnificent Dassault Falcon7X aircraft, flying subsonically at 48,000 ft. For us, the splendid totality lasted 4 min, while the pilot of a commercial Airbus, much lower down, was circling to allow all his passengers to enjoy this moment. This is an experience I would wish for all my readers! (Fig. 7.1)

But what about astronomers? Will they be able to do without eclipses if they wish to predict how the light and particles thrown into space by the Sun control the climate and life itself on our planet?

[1] http://xjubier.free.fr/site_pages/Solar_Eclipses.html



P. Léna, *Racing the Moon's Shadow with Concorde 001*, Astronomers' Universe,
DOI 10.1007/978-3-319-21729-1_7

**Fig. 7.1** The total solar eclipse of 20 March 2015 observed from a Falcon7X jet above the Atlantic, looking south-east. During totality, the Sun's elevation was 19° above the horizon. The corona glows (*top*) with a faint spurious reflection from the window. At this latitude, the umbra is elliptical in shape, with dimensions of about 300 km (in the SE-NW direction) by 90 km (SW-NE). National Center for Atmospheric Research, © UCAR

## An Apparently Simple, but Stubborn Question

In June 1973, the Sara villagers of Takamala, a tiny settlement on the banks of the river Chari in the north of the Central African Republic, had the following discussion with three French ethnologists: "Why did you choose us, the people of Takamala? Why did you bring this eclipse on here, and not in your own country? The Americans have been to the Moon ... You and your Concorde, you have hidden the Sun." Whether it was humour or irony, they were much closer to the truth than they imagined, for since 1937, astronomers have indeed been able to produce artificial eclipses using an instrument known as a coronagraph, invented by the French astronomer Bernard Lyot (see the appendix). However, this invention never completely removed the interest of eclipse expeditions because, from the ground, it can only be used to photograph the lower regions of the corona, which are sufficiently bright.

The two decades between 1960 and 1980 were important years for scientific eclipse expeditions. In this narrow band of solar atmosphere in which the temperature suddenly goes up, scientists wanted to find out what mechanism could carry such a large amount of energy up to these altitudes in order to heat the gases to more than a million degrees. If the acoustic waves studied by two of the instruments aboard Concorde were not in fact the prime cause of this heating, then a better answer to the question still remained unavailable. But could progress be made with

**Fig. 7.2** Just before take-off in 1964, the balloon that would carry Coronascope II from the High Altitude Observatory in Colorado up into the stratosphere. National Center for Atmospheric Research, © UCAR

artificial eclipses? After leaving the US Navy, Jack Eddy, encountered earlier in this story at the foot of the Rocky Mountains, became interested in this issue in 1960 when he was preparing his doctorate. Above Wisconsin, 25 km up in the stratosphere, his unmanned sounding balloon carried an automatic telescope called Coronascope I, designed to view the Sun. Direct light was blocked out by an occulting disk located some 2 m from the telescope objective, thus forming an external disk Lyot coronagraph.[2] Without measuring the corona directly during the first flight, Jack was able to show that, at this altitude, the blue glare of the sky was suitably reduced to less than ten billionths of the Sun's brightness. Four years later, Coronascope II obtained magnificent pictures of the outer corona using this method (Fig. 7.2).

However, this tremendous achievement was pipped at the post because, just a year earlier, a coronagraph carried aboard a US Navy sounding rocket had obtained the first non-eclipse image of the outer corona. In the south Algerian desert, a French group led by professor Jacques Blamont and my friend Roger Bonnet immediately repeated this feat by launching another sounding rocket, this time French, called *Véronique*. It, too, was equipped with a coronagraph, but unfortunately, the parachute that should have broken its fall did not work properly and the photos could not be recovered.[3]

---

[2] http://www.leif.org/EOS/Eddy/1961_PhD-thesis.pdf

[3] R. Bonnet: *Les premières expériences françaises de physique solaire dans l'espace*, ed. by B. Schürmann, European Space Agency, ESA SP-472 (2001) p. 59.

From this time, carried by balloons,[4] rockets, and especially spaceborne observatories to altitudes where the sky background is truly dark, coronagraphs have been able to refine observations of the outer corona by procuring a permanent artificial eclipse. In 1995, a joint project by Europe and the United States launched the Solar and Heliospheric Observatory (SOHO) into space with several coronagraphs aboard. It has been 20 years since this launch and, even while I write, SOHO's instruments are still sending their observations back to Earth. Even today, I can watch yesterday's film of solar flares on the Internet, as monitored by SOHO (Figs. 7.3 and 7.4).[5]

**Fig. 7.3** Extract from a film made by the Large Angle and Spectrometric Coronagraph (LASCO) aboard the space-based SOHO observatory, situated between the Earth and the Sun. The central disk blocks out the Sun (whose size is shown by the *white circle*) and the lower corona. The light radiation is at the visible wavelengths detected by the human eye. Matter ejected by the Sun to form the outer corona is visible against the darker background in which the stars can be seen to shine. © SOHO (ESA & NASA)

[4] R.M. MacQueen: http://www.opticsinfobase.org/ao/abstract.cfm?uri=ao-7-6-1149

[5] http://soho.esac.esa.int/

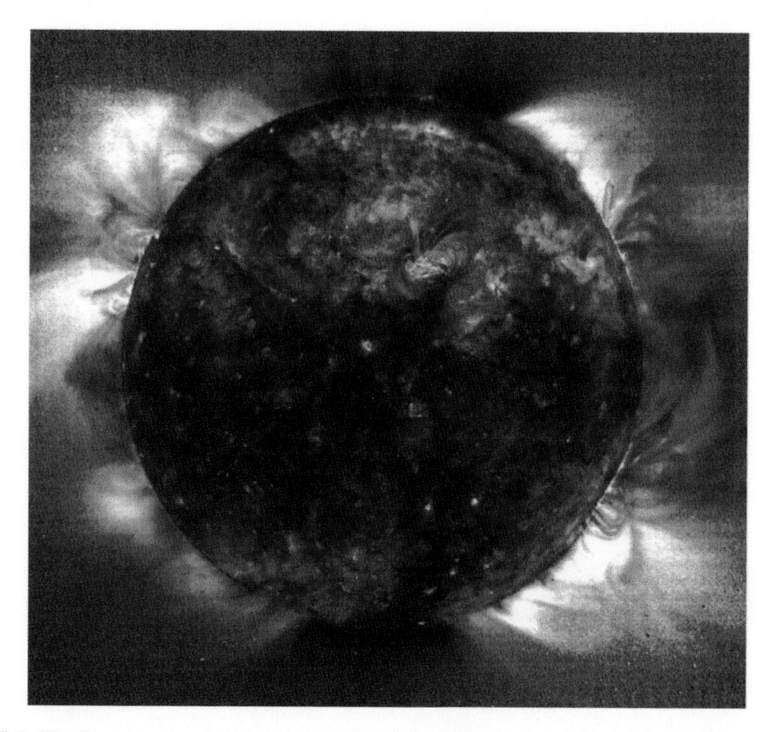

**Fig. 7.4** The Sun photographed in extreme ultraviolet light by the spaceborne observatory SOHO. At this wavelength of light, a coronagraph is no longer necessary, since the near corona is much brighter than the Sun's disk itself, being at a higher temperature. The non-uniformity of the temperature in the upper layers of the Sun is clearly visible here. © SOHO (ESA & NASA)

So in 2014, four decades after Concorde landed at *Le Bourget* and the rocket *Véronique* came to ground in the Grand Erg of the Saharan Desert, we still do not know the whole story about coronal heating, even though certain features have been pinned down. Regarding the rings or shells of dust, there is still doubt over their existence, despite the fact that the SOHO observatory has provided films of thousands of comets which graze the Sun and dissipate their matter, rather as happens around the star Beta Pic.

## Achieving the Impossible

Scientific research requires endless patience: while reality stimulates our desires, it can also resist them. It is also a long-running exercise in imagining the impossible. So let us formulate an observation and a question. First the observation: the Moon, so far away, tends to block out the Sun in an ideal way only too infrequently, while the mask of a coronagraph fixed only a few metres from a spaceborne telescope still tends to produce too many splashes of light which drown the image the telescope makes of this narrow zone where the coronal heating gets underway. And now the question: would it be possible to position the mask, a kind of artificial Moon, far enough in front of the satellite carrying the telescope to ensure that no more splashes of light could find their way into the image? Calculations show that the separation would only need to be a few 100 m. However, fixing the mask in such a way that it remains strictly at the same position in front of the telescope, perhaps held on a long mast carried by the satellite, is not a realistic option.

A solution would be two satellites flying in formation, the first carrying the mask and the second the telescope. They would then move in concert through space like Columbus' caravels on the ocean, using tiny gas jets to hold them precisely aligned with one another and with the direction of the Sun. They could thus orbit together for a decade or more, while flying in formation in this way. Indeed, I have before me the study showing that such a mission, known as DynaMICCS, is now a real possibility (Fig. 7.5).[6]

**Fig. 7.5** Project for permanent monitoring of the lower corona by a spaceborne instrument. Two satellites fly in precise formation, one (*right*) carrying a mask designed to occult the light from the solar disk, and the other (*left*), a few 100 m away, carrying the telescope and camera collecting images of the corona. DynaMICCS mission project put forward in 2009 at the European Space Agency. © Turck-Chièze et al., The DynaMICCS perspective. Experimental Astronomy, V 23 Issue 3, pp 1017–1056

[6] The acronym DynaMICCS stands for 'Dynamics and magnetism from the inner core to the corona of the Sun'.

I have no doubt that this mission will go ahead in the near future, and not only for the pleasure of understanding the corona. For we depend on the Sun, our star, which inspired the poet Edmond Rostand to write[7]:

Ô Soleil! toi sans qui les choses
Ne seraient que ce qu'elles sont!

He didn't know how right he was. Not realising in those days the effect of the Sun and its whims on the Earth's climate, radio communications, the health of astronauts in space, and on the whole of life. He would never have imagined that, at the beginning of the twenty-first century, we would be speaking of space weather, that is, the weather out in space, as something we really need to know about today. But it turns out that this weather depends to a certain extent on what is happening in the narrow layer of atmosphere around our star, as revealed to us by either true or artificial eclipses, something we will doubtless continue to admire and study for years to come.

[7] Oh Sun, without whom things
Would only be what they are!

# Epilogue

Strangely enough, I just cannot recall what triggered the idea of following an eclipse aboard Concorde, much as I would search my memory and even my notes. Using their telescopes, the Galileo I and Caravelle aircraft on which I had flown previously had brought me close to the stars. I vaguely recollect reading in a popular American astronomy magazine,[1] back in 1969, a suggestion by American colleagues who had sought to use the supersonic military aircraft SR-71 Blackbird, a spy plane, hence top secret, to follow the eclipse of 7 March 1970 on the east coast of the United States. But in 1970, the request for authorisation was rejected by the US Department of Defence. If they had got permission in 1973, they would have been able to fly above us at an altitude of more than 20,000 m, thus obtaining two good hours of totality. However, they would only have had a tiny space in which to set up one instrument and only the two pilots would have witnessed the spectacular event. But they would have got the record! So maybe this little article had worked its way into my mind during the two years leading up to the lunch in Toulouse. I just don't know. The origins of our adventure will remain forever shrouded in mystery.

There was no follow-up to the 1973 flight, despite our efforts in 1974 to try to repeat it with the British Concorde 002 prototype during the eclipse of 1976 above the Indian Ocean; with John Beckman, we had suggested a fascinating photographic programme based on the excellent images obtained by Jean Bégot. But things are doubtless better that way, since our flight over Africa remains unique to this day. In any case, Air France didn't forget about it. When people began talking about the total eclipse of the Sun of 11 August 1999 that would cross France from east to west, the airline decided to charter one of its Concorde, the Fox-trot Charlie F-BVFC, to take wealthy tourists to Ireland to admire the spectacle. The intrepid astronomer Audouin Dollfus, then aged 75, was invited aboard. Audouin was the son of Charles Dollfus, an aerostation pioneer whose little book *Histoire de l'aviation*, published on poor quality paper in the middle of World War II, had

[1] R. Mercier, J.M. Pasachoff: Ninety minutes of totality! In: Sky and Telescope, January 1969.



P. Léna, *Racing the Moon's Shadow with Concorde 001*, Astronomers' Universe,
DOI 10.1007/978-3-319-21729-1

been one of my childhood favourites. Audouin had been a student of Bernard Lyot, the inventor of the coronagraph mentioned above, and had bravely made the first manned balloon flight into the stratosphere on French territory, reaching an altitude of 14,000 m. So here he was aboard Fox-trot Charlie, flown by the pilots Jean Prunin and Eric Célérier. With my young colleague Vincent Coudé du Foresto, they delighted the passengers with details of the eclipse which turned out to be rather hard to see through the side portholes of the cabin. Totality was rather brief, a mere 5 min.

The music of Paul Verlaine still rings in my ears when I recall that beautiful morning in Las Palmas when we climbed aboard the great white bird that would carry us to the stars, in fact, toward our own star, the Sun, and our natural satellite, the Moon. The poet whispers[2]:

> Dis, qu'as-tu fait, toi que voilà,
> De ta jeunesse?

Somewhere above the North Atlantic, from the cockpit of the Air France Concorde F-BVFC, specially chartered for a commercial tourist flight for a rendezvous with the Moon's shadow during the total eclipse of 11 August 1999, one that many in France would be able to observe. © V. Coudé du Foresto

While writing this story, I thought of all those who, in the prime of life, are already haunted by the poet's question and dream of their life to come. I don't know how I would answer myself, if not by recounting my own story in these few pages, just as I experienced it with my companions in Las Palmas when Concorde 001 carried us up to "catch a shadow as our only prey",[3] for this is how André

[2] Tell me, what have you done, you there,
With your youth?

[3] A play on the French expression '*lâcher la proie pour l'ombre*'.

Turcat, who recently celebrated his ninetieth birthday, now qualifies our historic flight. And what better example than André Turcat to show that age is all in the mind?

Exactly 40 years after our absolute record for observation of a total eclipse of the Sun—74 min of darkness—eight members of that flight met at the *Musée de l'air et de l'espace* in *Le Bourget* on 29 June 2013, under the 001 prototype, in the presence of André Turcat (*centre*). *Left to right*: Michel Rétif (test flight navigation mechanic) and the science team John Beckman, Donald Liebenberg, Alain Soufflot, Paul Wraight, Pierre Léna, and Donald Hall. © P. Léna

# Appendix A: A Little Eclipse Dictionary

**Fig. A.1** With these beautiful stamps issued in 1973, the post office of the Republic of Senegal explained the mechanism underlying a total or partial eclipse of the Sun. The path of totality did not enter Senegal so the eclipse was only partial there. Document kindly provided by Henri Aubry

## A.1 Why Are Total Eclipses of the Sun so Rare on Earth?

To predict the occurrence and details of a solar eclipse, one must take into account the annual motion of the Earth around the Sun relative to the stars, the monthly motion of the Moon around the Earth, and the diurnal rotation of the Earth about its own axis. These three motions, or revolutions, have different and quite independent periods, the first being around a year, the second roughly a month, and the third a day. The motion of the Earth around the Sun takes place in a plane called the ecliptic plane, whereas the motion of the Moon around the Earth actually takes place in a different plane, and it is this that explains why eclipses are so rare.



P. Léna, *Racing the Moon's Shadow with Concorde 001*, Astronomers' Universe,
DOI 10.1007/978-3-319-21729-1

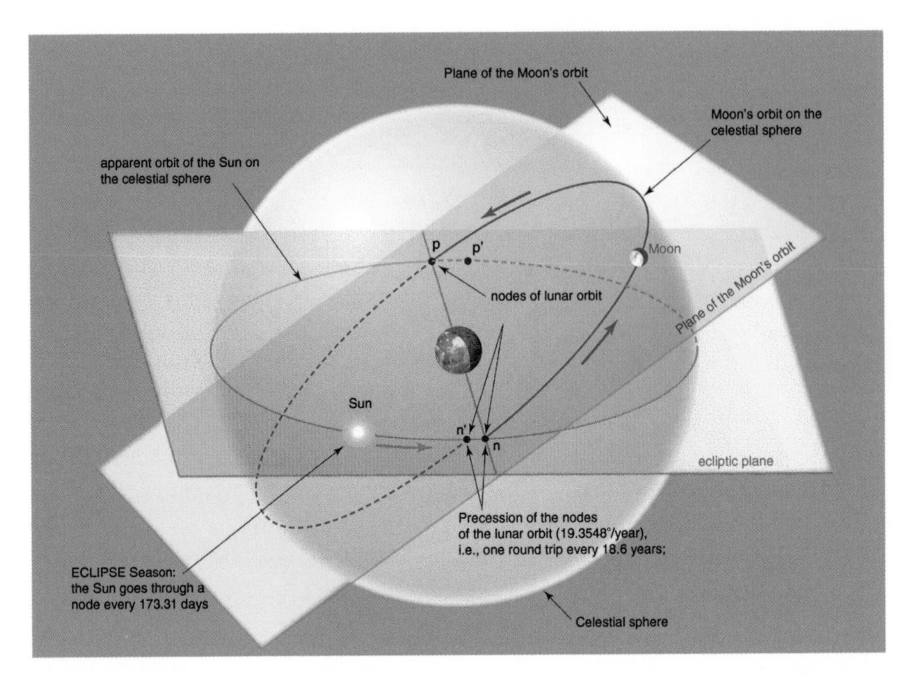


**Fig. A.2** Positions of the ecliptic plane (*yellow*) and the plane of the Moon's orbit (*blue*) relative to the Earth (*centre*). Seen from Earth, the Sun (*yellow*) follows a certain path (the ecliptic) on the celestial sphere (*red*), and the Moon likewise (*blue*). The straight line **pn** is the line of nodes at a given time, noting that the points **n** and **p** move to points **n'** and **p'** after a certain time. © IMCCE/P. Rocher

The orbits of the Earth around the Sun on the one hand, and of the Moon around the Earth on the other, are not circular. In fact they are slightly oval-shaped curves called ellipses. The distance from the Earth to the Sun, and also the distance from the Earth to the Moon, thus change during one revolution around one of these ellipses. As a consequence, the apparent diameters of the two bodies as viewed from the Earth change slightly depending on whether the Moon is at its nearest point to the Earth (perigee), and whether the Earth is at its nearest point to the Sun (perihelion).

Two consecutive full moons are separated by 29.531 days (the lunar month or synodic period of the Moon, corresponding to 29 days 12 h 44 min and 3 s). At the full moon, the Moon lies in the exact opposite direction to the Sun, as viewed from the Earth. The same time separates two new moons, when the Sun and Moon lie in about the same direction. This duration results from the combined effects of the first two motions mentioned above.

One might expect the Moon to come between the Earth and Sun at each new moon, thereby producing a total eclipse. This would indeed be the case if the plane containing the orbit of the Moon was the same as the ecliptic plane. Each total eclipse would then be separated from the last by a period of 29.531 days. But in fact, these two planes are distinct. There is a small angle between them, just $5°8'43''$. The Moon's orbit therefore cuts the ecliptic plane at two diametrically opposite points called nodes. The line of nodes is the straight line joining them. If as viewed from the Earth the line of nodes coincides with the direction of the Sun at the moment when the Moon passes through this node, there will then be an eclipse of the Sun. This eclipse will be partial or total depending on whether the coincidence is approximate or exact, producing therefore differing degrees of overlap between the two disks. The occurrence of an eclipse of the Sun thus depends now on the orientation of the line of nodes relative to the Sun.

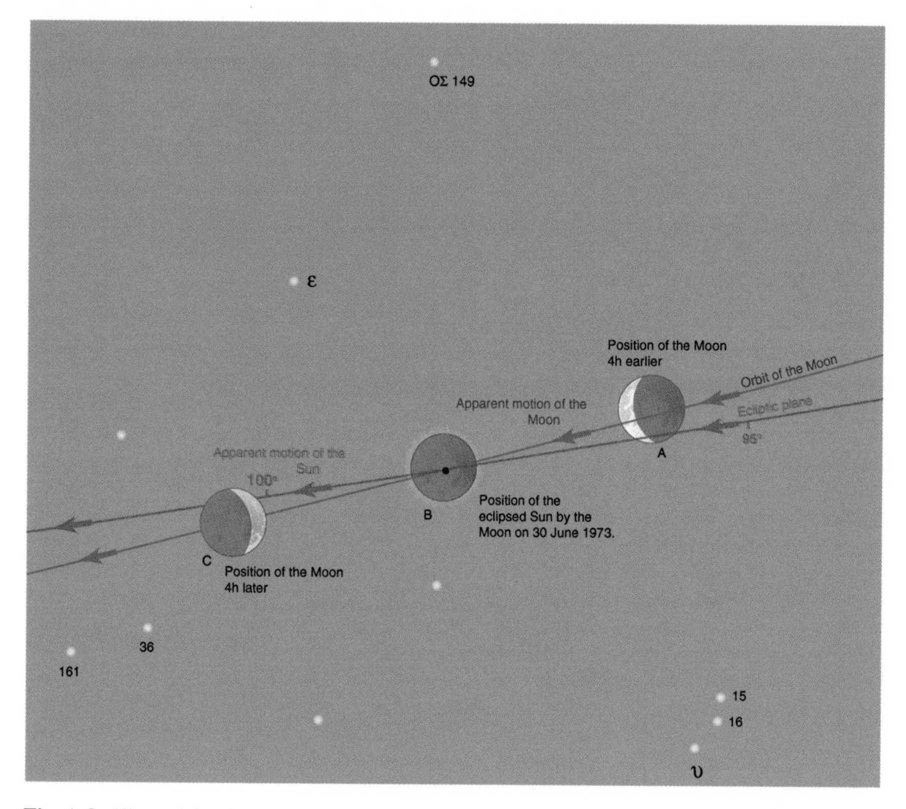


**Fig. A.3** View of the sky as seen from West Africa during totality on 30 June 1973. The position of the eclipsed Sun is shown (to scale). It coincides at this instant with the position of the Moon, at an ecliptic longitude close to an angle of 100°. As time goes by and for an Earth-based observer, the Sun moves slowly along the ecliptic (*red curve*) relative to the stars (on 30 June, it was here in the constellation of Gemini). Likewise the Moon moves relative to the stars (*blue curve*), but more quickly. Its path is shown, together with its position 4 h before and 4 h after totality. The stars lying in directions close to that of the Sun are identified. © P. Léna

The three motions mentioned above are governed by the rules of celestial mechanics, and so too is the motion of the line of nodes. Under the effects of gravity, the direction of this line rotates slowly relative to the direction of the Sun, making a complete turn every 18.61 years. This fourth motion must therefore be taken into account when predicting eclipses.

Figure A.3 sums up the situation, as seen from the Earth. The plane of the ecliptic projects onto the celestial sphere, where it traces out a virtual curve called the ecliptic. This curve crosses the constellations whose names are associated with the signs of the zodiac. At any given time, the direction of the centre of the Sun is located at a point on the ecliptic (the longitude of the Sun). This point moves during the year. It passes through one of the two nodes every 173.31 days (roughly twice a year).

Still viewed from the Earth, the plane of the Moon's orbit also projects onto the celestial sphere, where it traces out another virtual curve. At any given time, one can always mark the latitude and longitude of the direction toward the centre of the Moon on this curve, and also the direction of the nodes, which always lie on the ecliptic.

The Moon travels round its orbit and goes through the same node (one of the two) on each round trip. The time elapsed between these two passages is 27.212 days (known as the draconitic period). It differs from the lunar month (29.531 days) discussed above, because it is independent of the motion of the Earth around the Sun, which is not the case for the phases of the Moon, like the new moon.

Finally, since the Sun and the Moon, when seen from the Earth, are not points but disks, extending over a certain region of the starry background, the size of these disks relative to their respective centres can be shown on the figure.

From what has been said so far, we can now understand what determines the occurrence of solar eclipses, whether total or partial, and their repetitions.

Consider again Fig. A.3. The exact time of the new moon is when the two longitudes, that of the Sun moving along the ecliptic and that of the Moon, are equal. This happens during each monthly lunar revolution. An eclipse of the Sun can thus occur close to this time, and one must take into account the sizes (roughly half a degree) of the solar and lunar disks:

- If the latitude (coordinate measured perpendicularly to the longitude) of the centre of the Moon is less than 1.42°, an eclipse will certainly occur.
- If the latitude of the centre of the Moon lies between 1.42° and 1.58°, an eclipse may occur.
- If the latitude of the centre of the Moon is greater than 1.58°, it will not occur.

These angles can be converted into the longitude of the centre of the Sun, by comparing with the longitude of the node of the lunar orbit. As the angle between the plane of the lunar orbit and the plane of the ecliptic is very small (about 5°, see above), the difference between these two longitudes corresponds to larger angles:

- If the difference between these longitudes is less than 15.665°, there will be an eclipse.
- If the difference lies between 15.665° and 17.375°, there may be an eclipse.
- If the difference is greater than 17.375°, there will not be an eclipse.

Depending on the differences in latitude, or longitude when the Moon passes close to the node, and depending on the apparent diameter of the two bodies at this moment of time, the eclipse of the Sun may be central (total, hybrid, or annular) or partial.

One may wonder whether the motions of the Earth around the Sun, the Moon around the Earth, and the line of nodes in the plane of the ecliptic can work together to produce eclipses on a more or less regular basis. In this context, it can be noted that a period equal to 242 times the draconitic period is almost exactly equal to 223 times the synodic period of the Moon. In other words, after $242 \times 27.2122208 = 223 \times 29.5305882 \approx 6585.32$ days, the centres of the Sun, the Earth, and the Moon will be found very close to the same positions and a new cycle can begin. This period of 6585.32117 days (or 18 years 11 days 8 h) is called the saros. One saros contains on average 42 solar eclipses, of which 14 will be partial and 28 central (total, hybrid, or annular). These numbers may vary slightly from one saros to another. The exeligmos is a period of three saros cycles, using the Greek term for 'revolution' or 'orbit', but the term also refers to the sinuous curve followed by a fleeing hare! Two total eclipses, e.g., with the diamond necklace effect, separated by three saros cycles, will occur at the same local time on Earth.[4] Saros cycles are numbered from the year 2000 BC, and the saros of the eclipse of 30 June 1973 is number 136.

Since the celestial motions are known with great accuracy and are reproduced identically over time, the date and time of each eclipse can be calculated. Many tools are available on the Internet to provide information, for example, at the site of the astronomer Xavier Jubier, or again the *Institut de mécanique céleste et de calcul des éphemerides*, from which the above has been adapted.[5]

## A.2 Why Do We Only Observe an Eclipse of the Sun from a Very Small Region of the Earth's Surface?

When we observe the shadow of a given object due to some source of light, there are two possible situations: the source may be a point of light or it may be extended in space. In the first case, the shadow will be perfectly clear, with a sharp boundary in going from the dark to the light zone. In the second case, there is an intermediate zone called the penumbra between the fully lit and totally dark zones. The penumbra is partially lit, that is, by only a part of the source. Since the Sun is of course not a point but extended, the shadow of the Moon on the screen formed by the Earth will thus contain a penumbra. Given the apparent size of the Sun (about half a degree), the size of the Moon and its distance from Earth, the diameter of the

[4] L. Robert Morris may be the first person to have superposed two images of totality taken at an interval of one exeligmos (three saros cycles). Sky and Telescope, August 1966.

[5] The French amateur astronomer Xavier Jubier has made available an accurate and user-friendly calculator for past and future eclipses on his website: http://xjubier.free.fr/site_pages/Solar_Eclipses.html and http://xjubier.free.fr/site_pages/astronomy/ephemerides.html. A rich source of information is also available at the site of the *Institut de mécanique céleste et de calcul des éphémé rides* (Paris Observatory): http://www.imcce.fr/fr/ephemerides/phenomenes/eclipses/soleil/chap01.php

shadow will be at most 269 km (when the Moon is at its closest point to the Earth), while the maximum diameter of the penumbra will be 3000 km. The eclipse is total when observed from a point on Earth located in the umbra (full shadow) and partial when it is located in the penumbra. There is no eclipse when viewing from outside the penumbra.

The shadow moves across the Earth's surface roughly from west to east due to the motion of the Moon on its orbit around the Earth, with a speed of some 3400 km an hour. However, the rotation of the Earth about its own axis, also from west to east, subtracts the Earth's rotational speed from this (1660 km an hour at the equator). The result of this difference of speeds is the resultant speed of the shadow, which lies between 3400 km an hour near the poles and 1700 km an hour at the equator.

In extraterrestrial space, the region where the Sun completely hides the Moon is a cone, the shadow cone. The present distance between the Moon and the Earth is such that the tip of this cone only just reaches the Earth's surface, masking the Sun completely. Spacecraft sent into space by humans and moving through the Solar System experience a total eclipse of the Sun whenever they cross this shadow cone. The other planets (Venus, Mars, etc.) each have their own shadow cone.

The position of the totality zone on the Earth's surface is determined by the eclipse conditions (longitudes and latitudes), which in turn determine the point where the centre of the umbra is located on the surface and the so-called centrality region swept out by the umbra.

## A.3 Why Was the Total Eclipse of 1973 Called the Eclipse of the Century?

A total eclipse of the Sun lasts longest when the umbra is as extensive as possible, that is, when the Earth is as far as possible from the Sun (aphelion), the Moon as close as possible to the Earth (perigee), and the Sun is at the zenith of the observation point, in which case the umbra is circular. The first condition fixes the date of such an eclipse, close to the summer solstice. The third condition fixes the point of observation, close to the tropic of Cancer at the summer solstice. Under such conditions, the speed of the shadow is 2196 km an hour and its diameter is 262 km, implying a duration of 7 min and 10 s. The calculation may need to be corrected somewhat, and the duration at a point further south, at only 5° from the Earth's equator, can reach an absolute maximum of 7 min and 30 s (the diameter of the shadow decreases, but the speed too, and the ratio of the two quantities turns out to be more favourable).

For an annular eclipse, the conditions are slightly less strict and quite opposite: the apparent diameter of the Sun's disk must be as big as possible (Earth at perihelion) and the diameter of the Moon's disk as small as possible (Moon at apogee). An annular eclipse can last as long as 12 min and 30 s.

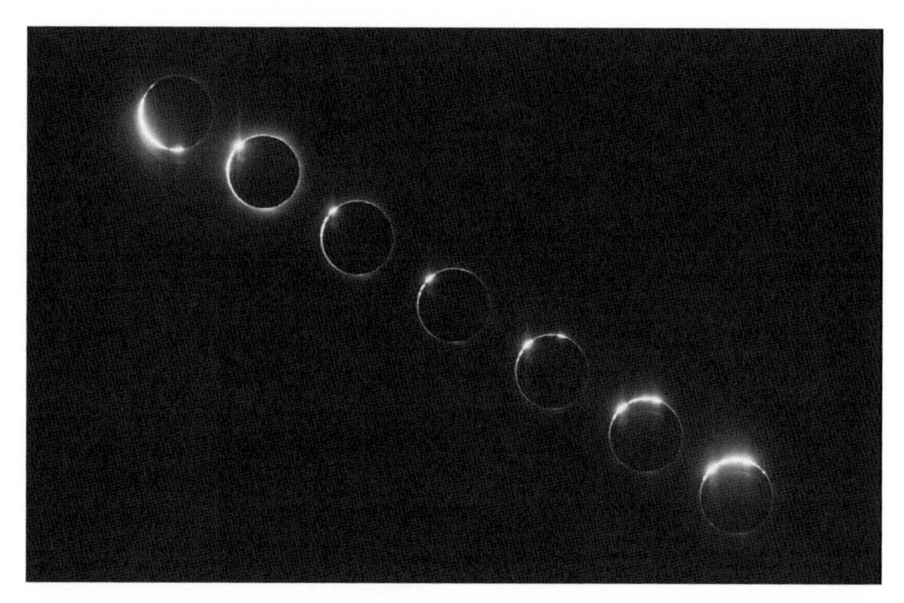

**Fig. A.4** Total solar eclipse of 3 November 2013, photographed at an altitude of 13,700 m from a Falcon aircraft above the Atlantic, 600 nautical miles to the south-east of Bermuda. For this eclipse, the distances from the Earth to the Sun and to the Moon are such that, at the moment of totality, there remains a very thin ring of chromospheric light. To get this sequence, the aircraft deliberately flew at right-angles to the line of centrality, whence totality only lasted for an extremely short time. © B. Cooper/LaunchPhotography.com

On 30 June 1973, the positions of the Earth, the Moon, and the Sun were such that, for a ground-based observer located not far from the equator, totality lasted for a maximum of 7 min and 4 s, very close to the absolute possible maximum.

# Appendix B: The Solar Corona

What is meant by the solar corona? The Sun is a ball of gas, essentially made up of the simplest atom, that is, hydrogen. This ball, an almost perfect sphere, holds itself together in a certain sense, like almost all celestial bodies, by the mutual gravitational attraction of the atoms making it up, which stops them dispersing into space. At the centre of the Sun, the gas is heated to several million degrees by nuclear reactions, but its temperature drops to about 4500° at the surface (the photosphere), which is what we see when we look at the Sun in the sky. The apparent size of the solar disk as viewed from the Earth is half a degree of angle (0.5°), corresponding to a diameter of 1,400,000 km.

But this apparent surface does not mark the outer edge of the Sun. Above it, even though the gas becomes more and more tenuous and hence transparent, the temperature begins to increase. In a few thousand kilometres, it reaches several million degrees. The cause of this sudden increase has still not been perfectly identified. At these temperatures, the hydrogen atoms break up into two electrically charged particles: protons with positive charge and the very light electrons with negative charge. Together these form what is known as a plasma. Furthermore, whereas the gas is very regularly distributed in successive layers as far out as the photosphere, higher up, it is organised into gigantic and magnificent shapes with much greater complexity, which continually vary as time goes by: arches, loops, jets of matter, and protuberances, all of which are grouped together as the lower corona (also known as the K corona). The reason for this sudden transition is understood: it is the effect of the Sun's magnetic field, which acts on the paths of the moving protons and electrons. Although the Sun which rises every morning looks quite the same to us, its corona is highly variable from 1 h to the next, 1 day to the next, and 1 year to the next. Due to its high temperature and the magnetic forces, the coronal gas, mainly composed of protons and electrons, cannot remain 'attached' to the Sun, and a part of it escapes to form what is known as the solar wind, which reaches the Earth and the other planets of the Solar System.

The visible light emitted by the K corona is a beautiful white colour. It is due to scattering of light from the solar disk on the electrons in the plasma. This very hot



P. Léna, *Racing the Moon's Shadow with Concorde 001*, Astronomers' Universe,
DOI 10.1007/978-3-319-21729-1

plasma also emits light, but at other wavelengths, either in the ultraviolet or X rays, or at radio frequencies. These wavelengths cannot be seen by the human eye, which is not sensitive to them, but they can be detected by ground-based or spaceborne telescopes equipped with suitable devices.

The coronal gas thus fills a huge bubble centered on the Sun. But it is not alone. Tiny solid dust particles, measuring around a thousandth of a millimetre across, are present throughout the Solar System. These have many different sources. Some have been around since the formation of the Sun and the planets about 5 billion years ago. Others are carried into the Sun's neighbourhood by comets which evaporate if they come too close, when their nucleus is heated up by the solar radiation. And yet others may result from collisions in which larger asteroids are broken into smaller parts. The chemical composition of these dusts is known: they are for the main part silicate grains, of the same kind as the rocks in Earth's mountains. Like the planets, they follow elliptical or circular orbits around the Sun under the effects of gravity. Their temperatures are higher when they are closer to the Sun, which illuminates and heats them. In contrast to the solar wind, they are mainly concentrated in the vicinity of the ecliptic plane, the one containing the orbits of the planets and many comets. As the plane of the ecliptic cuts through the constellations of the zodiac, this dusty region is known as the zodiacal cloud. The cloud itself is extremely tenuous, but when a dust particle is picked up by the Earth's gravity and encounters our atmosphere, it produces a tiny shooting star. Each year, several tens of thousands of tonnes of these dusts are consumed in this way. These tiny grains, lit up by the Sun, scatter the light in all directions, just like any grain of dust illuminated by a ray of light and forming a bright point in the darkness. At night, 2 or 3 h after sunset and when the sky is exceptionally clear, one can make out a sort of weakly lit oval, spread along the plane of the ecliptic, with an intensity that falls off as one observes further and further from the Sun. This is known as the zodiacal light. It is precisely the light scattered in our direction by these grains. This light is very faint indeed: at 90° from the Sun, it is ten-thousandth of a billionth ($10^{-13}$) of the brightness of the disk, reaching one billionth of that at 1° from the solar limb. Note in passing the extraordinary adaptive capacity of the human eye to different light levels, since it can detect all such levels from ordinary daylight to the zodiacal light.

**Fig. B.1** On this very wide field photograph taken from the site of the Very Large Telescope (VLT) in Cerro Paranal, Chile, the zodiacal light appears as a faint oval-shaped glow of light, almost at right-angles to the horizon and extending over several tens of degrees. *On the left*, the Milky Way stands above the 8.2 m telescopes. © Y. Beletsky (LCO)/ESO

The scattering electrons in the lower corona reflect very little visible light, that is, light visible to human eyes, with wavelengths lying between the red and the violet: at 1° from the solar limb, that is, at a distance from the limb equal to twice the diameter of the disk, the brightness of the light from the lower corona is of the order of one billionth ($10^{-9}$) of the brightness of the disk itself. From high up in the mountains, when the sky is very clear, the blue sky is a thousand times brighter due to the Earth's atmosphere, which scatters light from the solar disk! It is not surprising, therefore, that we are unable to observe a corona around the disk under such conditions. However, during a total eclipse of the Sun, the Earth's atmosphere is no longer illuminated and the sky background can become a thousand times less bright than the blue mountain sky, whence it will be comparable in brightness to the lower corona. The latter can then be made out and photographed above the solar limb. Moving further away from the limb, the lower corona becomes still less bright: 5° from the limb, it goes from a billionth ($10^{-9}$) to a thousand times less ($10^{-12}$), whereupon it is no longer observable.

An Earth-based observer receiving light emitted by the lower corona also receives light scattered by dust in the zodiacal cloud, which will be seen in cross-section: this is the F corona. At roughly 1° from the solar limb, the two contributions (K and F) are approximately equal, but the second soon dominates over the first when we view further away from the limb: whereas at 5° from the limb, the brightness of the K corona would be a thousandth of a billionth, that of the F corona, owing to the zodiacal cloud, is hundred times brighter, so only it would then be visible.

Apart from the solar light they scatter, zodiacal dust grains are also heated up by this radiation. When they are at a similar distance from the Sun to the Earth, they are also at a comparable temperature to the Earth, namely about 0 °C. But when they are much closer to the Sun, their temperature increases to the point where they are vaporised, at some 1500 °C, a temperature they do reach close to the Sun, when located within 3–4° of the limb. If we view regions within a few degrees of this limb, it should therefore be possible to observe the emission of infrared light from these heated grains. Viewing at higher angles, we will observe weaker emissions, indicating a lower temperature (Fig. B.2).

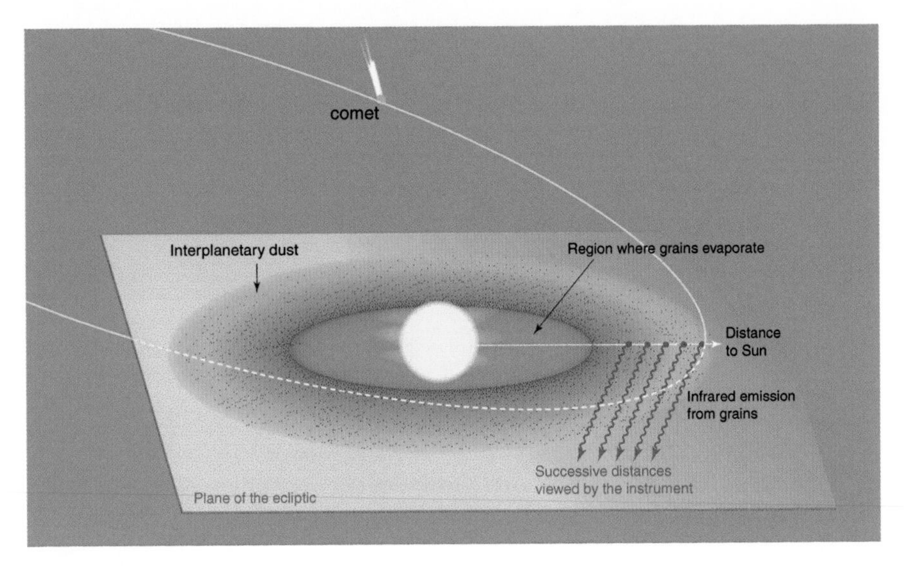


**Fig. B.2** Hypothetical circumsolar ring of interplanetary dust, target of infrared observations during the 1973 eclipse. © P. Léna

The infrared emissions from the zodiacal cloud are very weak, but they can make it very hard to carry out astronomical observations, even from space. This is the case when we try to measure the radiation coming from the earliest epochs of the Universe, when the galaxies were just forming, or even earlier. The magnificent Planck cosmology mission, launched into space by the Ariane rocket in 2009, encountered this difficulty, and astrophysicists required a great deal of ingenuity to subtract this undesirable light contribution due to the zodiacal cloud from their measurements.[6]

[6] http://public.planck.fr/resultats/228-planck-et-l-emission-zodiacale

# Appendix C: Invention of the Artificial Eclipse

Bernard Lyot, born in 1897, was a particularly good student. The scientific notebooks he so carefully drew up himself when, as an adolescent, he was already carrying out his own experiments, could inspire many teachers who would like to give their pupils a taste for science[7] (Fig. C.1).

When he became an astronomer at the Paris Observatory, he was fascinated by the Sun, which he observed from the world-renowned Pic du Midi Observatory, inaugurated in the Pyrénées in 1882.[8] Lyot wanted to be able to observe the solar corona every day. At the age of 30, in order to refute the claim by the German physicist Hans Kienle who said in 1929 that[9] "tests prove without doubt that it is impossible to photograph the solar corona during daytime", Lyot invented a subtle instrument which immediately made him famous: the coronagraph.

[7] B. Lyot: *Cahier d'expériences* (hand written), kindly communicated to the author by Gérard Lyot, Mireille Hibon-Hartman, and the Bernard Lyot Club.

[8] http://www.imcce.fr/en/observateur/campagnes_obs/phemu03/Promenade/pages5/545.html

[9] E. Davoust: *Le coronographe de Bernard Lyot au Pic du Midi*, at http://www.bibnum.education.fr/files/lyot-analyse-43.pdf



P. Léna, *Racing the Moon's Shadow with Concorde 001*, Astronomers' Universe,
DOI 10.1007/978-3-319-21729-1

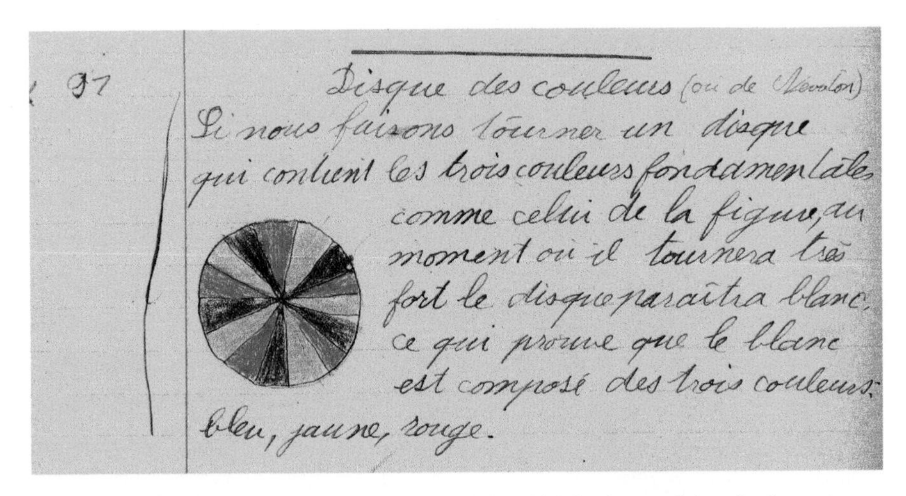

37

Disque des couleurs (ou de Newton)
Si nous faisons tourner un disque
qui contient les trois couleurs fondamentales
comme celui de la figure, au
moment où il tournera très
fort le disque paraîtra blanc,
ce qui prouve que le blanc
est composé des trois couleurs:
bleu, jaune, rouge.

**Fig. C.1** As a child, Bernard Lyot kept a notebook in which he drew and described experiments he carried out himself at the age of 13. Here he describes the experiment with Newton's disk. When rotated quickly, it shows how white light is a synthesis of all the colours of the rainbow. Courtesy of Pierre Léna

To hide the Sun, he inserted an opaque disk on the solar image formed in his telescope, hoping that this could imitate the role played by the Moon in a genuine eclipse. This simple idea failed because light was scattered in all directions from the edges of his disk. The scattered light then wiped out the weak glow of the inner corona, millions of times fainter than the disk, on the resulting photograph. With great care and with a cleverly placed second disk, Lyot managed to eliminate a large part of the unwanted light. Behind the double disk of his coronagraph, with no further need for a real eclipse to block out the unduly bright solar disk, he was then able to photograph at his leisure the chromosphere and the lower corona, regions where we observe the violent movements of solar matter called protuberances. At the Pic du Midi Observatory where he set up the coronagraph, Lyot produced unprecedented film sequences which showed solar flares as they happened. In 1952, Bernard Lyot went to Khartoum on the Nile in Egypt to observe a total eclipse of the Sun. He would never see France again, for he died of exhaustion a few weeks later in Cairo while examining his photos, which the Egyptian government had forbidden him to take back to Paris. In 1957, the documentary film *Flammes du Soleil* was made by the Paris Observatory in homage to Bernard Lyot. It achieved a huge success when projected in cinemas.[10]

[10] The film *Flammes du Soleil* (1957) is available from http://videotheque.cnrs.fr/index.php?ulraction=doc&id_doc=1348

**Fig. C.2** In 1939, Bernard Lyot, inventor of the coronagraph, observing the Sun with his instrument at the Pic du Midi Observatory, just before being accepted into the French Academy of sciences. Taken from the film *Flammes du Soleil*. © Joseph Leclerc /Observatoire de Paris

The Lyot coronagraph, which thus creates an artificial eclipse, was soon available in all observatories involved in solar observation. This instrument led to great progress in understanding the lower corona. However, the faint luminous envelope of the more distant corona still escaped observation, being not just a million times, but a billion times fainter than the Sun's disk. So not only do the ultimate residues of scattered light preclude observation of this distant corona, but even from a high mountain top, the glow of the Earth's atmosphere lit up by the Sun—the blue sky—produces a light background that completely masks this corona. The only solution is to go higher than the Pic du Midi. But where? Why not into space, from where we observe the corona today (see Chap. 7)?

# Appendix D: Concorde

This aircraft stands out in the second half of the twentieth century, not just as an extraordinary technical and aeronautical achievement, but also as an exceptionally daring human and commercial adventure.[11] Even though the plane was eventually withdrawn from service in 2003 after transporting thousands of passengers, these unparalleled successes gave the two countries involved in its development, France and the United Kingdom, the technical supremacy which allowed Europe, with the company Airbus, to become the leading supplier of commercial airliners in the world.

The possibility of carrying a hundred passengers safely at Mach 2, covering more than 6000 km without an airstrip available for rerouting during a large part of the flight, meant solving a host of novel problems. This is better appreciated by noting that, in the mid-1960s when the decision was taken to design Concorde, only one absolutely secret plane, the Lockheed SR-71 of the US Air Force, was even capable of cruising for long periods in supersonic flight, and then for a crew of just two people.

[11] See the books by André Turcat and Henri Ziegler in the bibliography.



P. Léna, *Racing the Moon's Shadow with Concorde 001*, Astronomers' Universe,
DOI 10.1007/978-3-319-21729-1

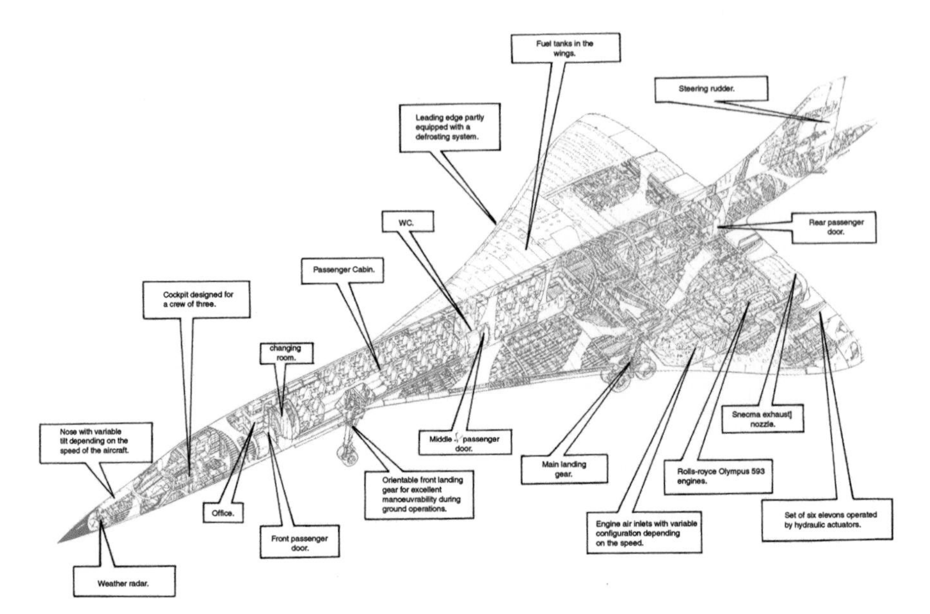


**Fig. D.1** Main features of the Concorde commercial airliner. © Aviation Magazine International and J. Pérard, DR

Let us list here a selection of these problems, together with the solution eventually found for them:

- Wings. These had to be designed with sufficient lift to be able to fly at low altitude (take-off and landing), and also in supersonic flight. The modified delta wing, or gothic wing, had enough capacity to carry the fuel reservoirs. Its low level of lift at low speeds meant the aircraft had to land in a nose up position.
- Engines. These had to be able to supply the necessary thrust, while operating with air injected at very variable speeds, something which has a major effect on combustion conditions in the jets: air injected while cruising at Mach 2 must become subsonic in the engine. The air intake was controlled by adjustable intake ramps. To avoid instability (jet pumping or compressor stall), these had to be permanently monitored.
- Afterburners. The afterburn or reheat increases the thrust in certain critical stages of the flight, namely, take-off and when crossing the sound barrier (from Mach 0.97 to Mach 1.7). The fuel is then injected into the engine exhaust fumes where it bursts into flame. This is very costly in terms of fuel and must be restricted to situations where it is absolutely essential, so not during cruising, for example.
- Flight control systems. The length of the plane makes it too difficult to use cables so the electric motors actuating the flaps and elevons are remote-controlled by electric circuits, for the first time aboard a commercial airliner.
- Brakes. The use of carbon disk brakes, equipped with a servo system, saves a considerable mass, avoids skidding, and reduces the stopping distance.

- Materials. The aluminium alloy used to build the plane must withstand a large increase in temperature in supersonic flight: more than 100 °C for the fuselage and the leading edge of the wings. Parts are milled directly from the bulk material.

All these innovations and many others were tried and tested on the ground and in flight, led in France by André Turcat and in the United Kingdom by his counterpart Brian Trubshaw, before they were adopted for the mass-produced plane. Many were then adopted into the design of subsonic Airbus planes. It was their well considered combination that allowed Concorde to meet the assigned specifications. The reader may refer to the many books detailing the Concorde venture.[12]

In 2014, now that the commercial airline fleets are solely subsonic the world over, many scientists and industrial groups continue to study the possibilities for supersonic commercial flight in the more or less distant future. For example, in 2011, the company EADS which makes Airbus presented a hypersonic aircraft project called Zero Emission HyperSonic Transportation (Zehst). Flying at Mach 4 at an altitude of 32,000 m, twice the altitude of Concorde, it could carry around a hundred passengers from Paris to Tokyo in an hour and a half! Maybe one day we will see this plane beat the record of 74 min of totality.

Always higher, always faster, always more air passengers? Is this reasonable? Is it sustainable, to use the modern term? The world has changed and humankind must learn new ways to inhabit this planet and exploit the resources here, with much greater thrift. Of course, we can still dream, but perhaps of different goals.

[12] http://fr.wikipedia.org/wiki/Concorde_(avion)

# Bibliography

André Turcat: *Concorde: essais et batailles*, Stock, Paris (1977)
Henri Ziegler: *La grande aventure de Concorde*, Grasset, Paris (1976)
André Turcat, Pierre Sparaco, Germain Champbost: *Une épopée française: les créateurs de l'aviation nouvelle, 1950–1960*, P. Galodé, Saint-Malo (2010)
Concorde [an ICBH Witness Seminar]: Kenneth Owen (Ed.), Institute of Contemporary British History, London (2002)
*Eclipse 73*: Thirty-five minute film in three languages (DVD), directed by Gilbert Dassonville, Jean Rösch, and Pierre Léna (scientific director), Cerimes, Vanves (1973)
*Soleil est mort: l'éclipse totale du 30 juin 1973*: Gérard Francillon et Patrick Menget (Eds.), Laboratoire d'éthnologie et de sociologie comparative, Nanterre (1979)
L. Robert Morris: Racing the Moon. In: *Ottawa Citizen Weekly*, 7 March 2004
L. Robert Morris: The day the Moon stood still. In: *Sky and Telescope*, May 2009, pp. 64–68
L. Robert Morris: *L'éclipse qui occulta le Titanic*. In: *L'Astronomie*, Société astronomique de France, April 2012, pp. 30–35
L. Robert Morris: *Biplan Voisin: l'éclipse de 1912, la naissance de l'astronomie en avion*. In: *L'Astronomie*, June 2013, pp. 26–29
Pierre Guillermier, Serge Koutchmy: *Eclipses totales: histoire, découvertes, observations*, Jean-Claude Pecker, Masson, Paris (1998)
Pål Brekke: *Le soleil, notre étoile*, translated by Jean-Claude Vial, CNRS Editions, Paris (2013)
Serge Brunier, Jean-Pierre Luminet: *Eclipses: les rendez-vous célestes*, Bordas, Paris (1999)
*La Chine des Ming et de Matteo Ricci (1552—1610): le premier dialogue des savoirs avec l'Europe*, Isabelle Landry-Deron (Ed.), Editions du Cerf/Institut Ricci, Paris (2013)



P. Léna, *Racing the Moon's Shadow with Concorde 001*, Astronomers' Universe,
DOI 10.1007/978-3-319-21729-1